Striking the Mother Lode in Science

"To me chaos is like a dream. It offers the possibility that, if you come over and play this game, you can strike the mother lode."
—Joseph Ford, a chaos researcher
at the Georgia Institute of Technology
(in Gleick 1987, p. 306)

Striking the Mother Lode in Science

The Importance of Age, Place, and Time

PAULA E. STEPHAN

SHARON G. LEVIN

New York   Oxford
OXFORD UNIVERSITY PRESS
1992

To our husbands and children:
Bill and David Amis;
Sandy, Allison, and Elliot Levin

Oxford University Press

Oxford   New York   Toronto
Delhi   Bombay   Calcutta   Madras   Karachi
Kuala Lumpur   Singapore   Hong Kong   Tokyo
Nairobi   Dar es Salaam   Cape Town
Melbourne   Auckland

and associated companies in
Berlin   Ibadan

Copyright © 1992 by Oxford University Press, Inc.

Published by Oxford University Press, Inc.,
200 Madison Avenue, New York, New York 10016

Oxford is a registered trademark of Oxford University Press

All rights reserved. No part of this publication may be reproduced,
stored in a retrieval system, or transmitted, in any form or by any means,
electronic, mechanical, photocopying, recording, or otherwise,
without the prior permission of Oxford University Press.

Library of Congress Cataloging-in-Publication Data
Stephan, Paula E.
Striking the mother lode in science : the importance of age, place, and
time / by Paula E. Stephan, Sharon G. Levin.
p. cm.   Includes bibliographical references and index.
ISBN 0-19-506405-4
1. Scientists—United States.   2. Creative ability in science-
-United States.   3. Ability, Influence of age.   I. Levin, Sharon G.
II. Title.
Q149.U5S73   1992
509.73—dc20   91-30827

9 8 7 6 5 4 3 2 1

Printed in the United States of America
on acid-free paper

Preface

This book is about scientific productivity and how events outside the immediate control of scientists shape their careers and achievements. We are most concerned with circumstances related to time. One such circumstance is aging. We began our study of productivity in science in an effort to answer the question, "Is science a young person's game?" Very early in this investigation we realized that age is not the only time dimension that affects the careers of scientists. Careers are also shaped by scientists' intellectual and economic environment, particularly at the time they are educated. Success in science depends, at least in part, on being in the right place at the right time, or, as we shall say, on "RPRT."

Some scientists are educated when rapid change in thought and practice is occurring in their field. They often become part of the change, ushering in new ways of thinking and looking at the world. Their careers prosper, especially the careers of those who are fortunate enough to have been students at the center of the intellectual movement. Other scientists come to their specialty at the wrong time, or study at a peripheral institution. All too rapidly they face the threat of knowledge obsolescence and the consequences obsolescence can have for success.

Economic events also play a major role in shaping careers. Scientists need specialized resources to be productive. Such resources are generally available only at research institutions. But research institutions are in the market for talent only when economic conditions permit. Some generations of scientists receive Ph.D.s only to find that jobs in the research sector are sparse. Compared to the careers of their colleagues who graduate during times of institutional growth and prosperity, their careers in science may languish for want of resources. This was certainly the predicament faced by the generation of physical scientists coming of age in the late 1960s and 1970s.

It is impossible to study the importance of age, time, and place without a framework for analysis. Hence, in an early chapter we discuss how science is done and why scientists do science. The "how" focuses on the special resources that individual scientists bring to the discovery process as well as the role of equipment and of colleagues in discovery. The "why" stresses the fascination of scientists with solving the puzzle and winning the game. To paraphrase a researcher in the field of chaos, science offers the possibility that if you play the game you can strike the mother lode.

The consequences of age, time, and place are of concern not only to scientists. They are, or should be, of concern to nonscientists as well. Science affects the quality of our lives and our competitive position in the world. Science affects our standard of living. In recent years the population of scientists in the United States has aged, a consequence of diminished job opportunities. Is an older scientific community holding back the United States? Has the aging of the scientific community contributed to the slowdown that has occurred in the rate of U.S. productivity? Has the problem been exacerbated by economic conditions that left a disproportionate number of young scientists without access to adequate resources? Has it been made worse in recent years as many talented young people have chosen careers outside science?

This book is not only about scientists. It is also about science policy. Although many of the issues we examine are outside the immediate control of an individual scientist, they are not outside the control of the larger society. Of course scientists age, but there is no reason why the scientific *community* must age. Neither is there good reason why access to scientific resources presently resembles a game of chance more than the rational enterprise it should be. Science in the United States faces serious problems today because of the absence of a long-term science policy.

ACKNOWLEDGMENTS

The research project upon which this book is based began more than a decade ago when Alan Fechter, then at the National Science Foundation and now Executive Director, Office of Scientific and Engineering Personnel at the National Research Council, called our attention to the fact that the U.S. scientific community was aging. Like ourselves, Alan is trained as an economist; he felt that the age-productivity question had been ignored too long by our profession. Throughout the course of our work, Alan has provided us with guidance and encouragement. We are grateful to him for setting us on the track, and for helping us keep a constant course. Alan never gave up, believing that funding was just around the corner and that the data, which proved difficult if not almost impossible to dislodge from the Washington bureaucracy, would soon be made available.

Funding for the project which formed the basis of this book came from the National Science Foundation, the Alfred P. Sloan Foundation, and the Exxon Education Foundation. Additional resources were provided by the Graduate School and the Gerontology Program of the University of Missouri-St. Louis, and the Georgia State University Research Office and College of Business Administration Research Council. All opinions expressed in the book are ours and should not be attributed to the granting institutions. Without the support of these agencies, we could never have done this research, particularly given the large expenditures incurred in assembling part of the database. Personnel at the granting institutions who were both helpful and patient during the period of our funding include Joel Barries and Morris Cobern, Division of Science Resource Studies, National Science Foundation; Eric Wanner at the Alfred P. Sloan Foundation; and Richard Johnson at the Exxon Education Foundation. We have also benefited from the support of the Policy Research Center, Georgia State University, and its director, Roy Bahl.

The research on which this book is based could never have been completed without the assistance of a number of persons. Only a few are mentioned here. Much of the data used in the book comes from matching the Survey of Doctorate Recipients with data from the Science Citation Index. The high quality of the match between the two is due in large part to the efforts of George Boyce and Susan Henn of the National Research Council, who were extremely generous with their time and always enthusiastic about the project. Kate Walker of the Institute for Scientific Information facilitated the exchange of information between the Institute and the National Research Council. Larry Pickett of the Office of Computing and Telecommunications at the University of Missouri-St. Louis deserves special thanks for processing the workfile tapes as quickly as possible.

In order to complete the econometric studies in physics and earth science reported in Chapter 8, it proved necessary to do case studies of the two fields. We owe a debt of gratitude to the many persons who were patient enough to work with us on these case studies and who refrained, at least most of the time, from suggesting that two economists had no business casting their nets so far afield. All told, over fifty persons were involved in this process. Some deserve a special thanks. First, it was Harriet Zuckerman who made us aware of the need for in-depth case studies. We wish to thank her for her early guidance and continued interest in the project. We also feel particularly fortunate to have had the help and support of John Rigden of the American Institute of Physics. John opened many doors for us, was always available to read drafts, and continually provided encouragement. Beverly Porter of the American Institute of Physics provided useful information for the physics case study and was readily available for comment, as was Spencer Weart, also of the American Institute of Physics. We also appreciate the assistance of Lily S. Recanati at the Institute. Other persons who were particularly helpful with the physics case study were Steve Manson and Steve Sigur. The completion of this case study is due in large part to Steve Sigur's willingness to tutor an economist in physics. In addition, we wish to thank Kirk Elifson who helped design the questionnaire, and the noted physicists who responded to our questionnaire.

For assistance with the earth science case study, we are indebted to several individuals. The late Bill Menard of the Scripps Institute of Oceanography discussed with us, at a very early stage, the primary focus of the project. William Sackett, Richard Montgomery Field Fellow at the American Geophysical Union, helped us locate respondents for the survey, and William Glen of the U.S. Geological Survey and *Eos* made useful comments. We are also grateful to the distinguished geoscientists who provided thoughtful responses to the survey and to Jon Stewart who shared insights gained from his own study of the plate tectonic revolution.

Lester Taylor discussed at length the estimating strategy used in Chapter 8 of this book. We are extremely indebted to him both for his help and for his continued enthusiasm for this project. Likewise, Frank Stafford helped with the theoretical model on which the empirical work is based, as did Rubin Saposnik.

Six Georgia State University graduate students worked as graduate research assistants on the project that formed the foundation of the book: Robert Eisenstadt, Stacy Kottman, Jan Luytjes, Vgee Ramiah, Cavery Bopaiah, and Lidija Polutnik. Vgee was particularly helpful in moving much of the Pullen Library to Paula's office. Lidija moved it back. The fact that the first three have long since received their Ph.D.s says something about the pace of this project. Kathy Groh, Leslie Sanazaro, and David Banks, students at the University of Missouri-St. Louis, also assisted in the project. Three persons who typed for the project deserve special thanks. At Georgia State University, Bee Hutchins typed the original proposal and continued typing on the project for several years. Marilyn King, also at Georgia State, picked up the project in midstream and helped see it to completion. Typing assistance at the University of Missouri-St. Louis was provided by Roxanne Viviano, who continued to work with us as the project progressed from report to book.

In 1988, having sent a final report to the granting agencies, we began to think about the possibilities of writing a book and approached several publishers. Most editors saw the work for what it was: an academic study, difficult to read and tentative in its policy implications. One editor, however, had a vision of what it could become. That editor was Herb Addison, of Oxford University Press, who believed that we could write a book that would not only interest economists, sociologists, and science policy makers, but could also, because of its style, attract a wider readership. Whether or not we achieve the latter will only be known after publication, but for his clear guidance and ability to call forth a writing style we did not know we possessed, we will be forever grateful to Herb Addison. He opened up a new world to us.

In the process of writing the book, we have sought the guidance and suggestions of many colleagues and acquaintances. The book is clearly richer because of their comments and suggestions. Despite their council, errors undoubtedly remain. We alone are responsible for these as well as for errors of omission. Those who were particularly helpful include Peter Ahrens, Mary Frank Fox, Ron Munson, Paul Roth, Mike Scherer, and Stephen Stephan. Thomas Hoisington, Marcia Mindell, and Ilene Wittels provided useful comments on the chapter on aging. The chapter on science policy benefited from discussion with Bill Rushing. Finally, but certainly not least, eight persons undertook the arduous and thankless task of reading a draft of the entire manuscript. Some read much of it twice. They are Lowell Hargens, Ralph LaRossa, Don Reitzes, John Rigden, Martin Sage, Rubin Saposnik, Steve Sigur, and Lester Taylor. We

Preface

would also like to thank Paul Bianchi, Bill and Laraine Tomassi, and Nancy Wiesner for their help.

Books are not written between the hours of 9:00 to 5:00. Instead, they take over your life, waking you up at 6:00 in the morning (if you succeeded in sleeping through the night) and consuming your weekends. To write a book requires the understanding and support of one's family, and in that we have been particularly fortunate. Both our husbands and children have understood the demands that the book has placed on our lives. They have listened to our concerns and anxieties; packed us off to "writing camp" each summer so that we could work, undisturbed, for several weeks; and generally understood the importance of completing the project. To them we owe a special thanks and to them we dedicate this book.

One family member has been particularly supportive in the writing of this book. Bill Amis, Paula's husband and professor emeritus of sociology at Georgia State, was able and willing to devote a particularly large amount of time to the project. It was always Bill who read the first draft, made suggestions, and provided the quick feedback we so much wanted. It was Bill, with Herb Addison, who made us believe that we could write a book in a style that escaped the academic verbiage so basic to our profession. Bill kept us going when we thought the end would never come, and it was Bill who did such a superb job of editing the manuscript that it went through the copy editing process at Oxford in record time. He also taught us more about hyphenation than we ever cared to know!

Finally, we are grateful that this research has given us the opportunity to work together during the past decade and has nurtured a friendship and a pattern of collaboration that began more than twenty years ago in Ann Arbor, Michigan, when we were graduate students in economics and spent long hours studying together for comprehensive exams. We know that we are fortunate to have each other.

Atlanta, GA   P. E. S.
St. Louis, MO   S. G. L.
1991

Contents

1. **Introduction, 3**
   Age, Place, and Time, 4
   Why These Issues Are of Concern, 5
   How the American Scientific Community Came to Be Older, 6
   What This Book Is Not About, 7
   The Plan of This Book, 7

2. **How Science Is Done; Why Science Is Done, 11**
   How Science Is Done, 11
   The Scientists, 11
   The Importance of Equipment, 13
   The Importance of Colleagues, 15
   The Role of Serendipity, 17
   Why Science Is Done, 17
   The Puzzle-solving Aspect, 18
   Recognition, 18
   Monetary Rewards, 20

3. **Why Age May Matter, 25**
   What Does Age Measure? 26
   Age and the Will to Do Science, 27
   Age and the Quest for Ribbon, 27
   Life's Passages, 30
   Age and the Quest for Gold, 33
   Age and the Lure of the Puzzle, 35
   Age and Cognitive Resources, 37
   Age and the Knowledge Base, 37
   Age and Mental Processes, 38
   Age and Creativity, 41
   Creativity as a Process, 41

      Creativity as a Product, 43
      A Head Start, a Nose for Success? 45
   Conclusion, 46

4. Age and Scientific Productivity: Must One Be Young to Do Great Things? 50

   Eminent Scientists, 50
      Lehman's Work, 50
      Dennis's Work, 54
      Nobel Laureates 55
   "Average" Scientists, 57
      Measuring Scientific Output, 57
      Collecting Data, 58
      Whose Productivity to Study? 59
      Cross-sectional Studies, 59
      Longitudinal Studies, 61
      The Productivity of Scientists Research (PSR) Study, 62

5. Does the Age Structure of Science Matter? 75

   Age Structure and the Rate of Scientific Discovery, 75
      Age and Resistance, 75
      Why Might Planck Be Right? 77
      Trivial Pursuits, 80
      Does Planck's Principle Hold? 80
   Science and Economic Growth, 84

6. The Importance of Place and Time, 90

   What Is RPRT? 91
   Job Market Conditions, 91
      The Importance of Location, 91
      The Job Market for New Ph.D.'s at Research Institutions, 93
      Supply and Demand for Scientists, 93
      From Boom to Bust: An Example of One Job Market Cycle, 94
      The Determinants of Current Location, 96
      Growth in the Field, 98
   Intellectual Climate, 99
      The Concept of Vintage, 99
      The Threat of Obsolescence, 101
      Growth and Change in Scientific Knowledge, 102
      The Importance of Fads in Science, 103
      Codification, 105
      How Scientists Cope with Change, 106
   Are Today's Scientists at Increased Risk? 109
   Summary, 111

7. Quality in Science, 115

What Is a Cohort? 115
Quality and Cohort, 116
The Selectivity Hypothesis, 117
The Secularization of Science, 123
The Brain Drain Hypothesis, 123
The Sociocultural Hypothesis, 126
Summary and Conclusions, 127

8. Age, Cohort, and Scientific Productivity: A Case Study, 132

What Are Pure Aging Effects? 133
Methodology, 134
Pooling and Identification, 134
Sample-selectivity Bias, 136
Individual-specific Fixed Effects, 137
Limited Dependent Variable, 138
Findings, 139
Areas of Science Studied, 139
Aging Results, 140
Cohort Results, 145
Summary, 152

9. Conclusion, 156

Age, 156
Right Place, Right Time, 158
Quality, 159
Cohort Differences, 160
Increased Competition, 160
The Consequences of Too Much Competition, 162
The Long-run Consequences of Impaired Productivity, 164
Stop-and-Go Funding, 165
Where Should We Go from Here? 165
Elements of a Rational Science Policy, 166
Who Should Pay? 166
More Applied Research, 167
What Should Be Funded? 167
Who Should Be Funded? 168
Reorganization at Universities, 168
Reform at the Elementary and Secondary Level, 169
Creative Options for Older Scientists, 170
A Conjunction of Need and Opportunity, 170

References, 175

Index, 185

Striking the Mother Lode in Science

1
Introduction

In 1983 *Esquire* magazine chose Robert Noyce as one of fifty Americans who "made the difference" in the twentieth century. Why? Turn on your computer, dial your cellular phone, play a CD, heat coffee in your microwave, watch a Patriot missile engage a Scud, and you'll have some appreciation of what Robert Noyce did. His name may not be a household word, but he ushered in the age of microelectronics with the invention of the integrated circuit in 1959.[1] At the time of the invention, Noyce had just turned 31 years old. Not only was he young; Noyce also had RPRT: He had been at the right place at the right time.

Robert Noyce grew up in Grinnell, Iowa, a town surrounded by rich farmland and the home to Grinnell College. Like his brothers, when the time came to go to college, Noyce enrolled at Grinnell, where he majored in physics. One physics professor at the college, Grant Gale, had known Noyce since Noyce's childhood. The summer before Noyce's senior year, Gale noticed a small item in the newspaper concerning the discovery of the transistor at Bell Laboratories. What caught Gale's attention was that one of the coinventors, John Bardeen, had gone to school with him. Wanting his students to be on the forefront, Gale wrote Bardeen, asking if he could have samples of transistors for his class. Not content to leave the matter there, Gale also wrote to the president of Bell Labs, himself a Grinnell alumnus. Two transistors arrived, and during Noyce's last semester at Grinnell he participated in the first course ever taught in solid-state electronics. Noyce was hooked. From there he went on to the Massachusetts Institute of Technology (where no course in solid-state electronics had yet to be offered), receiving a Ph.D. in 1953. For Robert Noyce, Grinnell College in 1949 turned out to be the right place at the right time.

Robert Noyce is only one example of a scientist who made a major contribution at a young age and who was fortunate enough to have RPRT. In this case, RPRT gave Noyce an intellectual edge over other physicists trained earlier or at a different place. RPRT can also provide scientists with a work environment that nurtures discovery. Such was the case for Walter Brattain, a coinventor of the transistor. When told that he had won the Nobel Prize in 1956, he is reported to have said: "Much of my good fortune came from being at the *right place*, at the *right time*, [emphasis added] and having the right sort of people to work with."[2] For Brattain and his coinventors, Bell Laboratories in the 1930s and 1940s provided RPRT.[3]

AGE, PLACE, AND TIME

Is there a right age to do science? Is there a right time and a right place? These are the questions we address in this book.

The belief that science is a young person's game is widespread, especially among scientists. The physicist P. A. M. Dirac was so much a believer that he was inspired to write, only half-jokingly:

> Age is, of course, a fever chill
> that every physicist must fear.
> He's better dead than living still
> when once he's past his thirtieth year.[4]

Noyce himself was convinced that the flashes of inspiration leading to discovery come mainly to the young, often those in their twenties. Other well-known believers in the importance of youth to scientific discovery include Albert Einstein, James Watson, and J. Robert Oppenheimer.

Scientists are also quick to understand the importance of RPRT. They know that place and time can be essential in science. Nobel laureates Linus Pauling and I. I. Rabi, as well as Oppenheimer, fully recognized this when they and other bright young scientists went to Europe in the mid-1920s to learn the new quantum mechanics. The subsequent rise to eminence of American science owes much to these young pioneers who understood the importance of RPRT.[5]

Good science necessitates knowledge of the frontier. Scientists who are particularly successful are often trained during a time when major innovations in thought or practice are occurring, or just after they have occurred. They become part of the change, carrying the revolutionary banner forward. Good science also requires up-to-date equipment and access to other scientists knowledgeable about what is going on in the field. Rarely does good science occur in isolation. Successful scientists usually work in fertile environments with access to substantial resources.

Many of the conditions that lead to RPRT are not specific to the individual but, rather, specific to a generation. This means that success in science depends, in part, on things outside the control of the individual scientist. When the scientific enterprise is growing rapidly, when governments are generously funding science, jobs in research centers are plentiful. Scientists wishing to pursue research careers thus have the opportunity to do so. *Job market conditions* are favorable for this generation of scientists. But when science is growing slowly, when government budgets for science are cut, it is difficult for new Ph.D.'s to obtain jobs in research centers. Indeed, a whole generation of scientists may find that they are frozen out of research centers, unable to obtain jobs conducive to research. They are, in a sense, sent to farm teams, never to be called up to the major leagues. For such a generation, job market conditions are unfavorable.

Some generations of scientists also have the good fortune to be educated at a time when the intellectual climate is ripe for significant breakthroughs. Such a generation constitutes a good *vintage*. Particularly blessed are members of the generation who study at a place where the changes are first introduced. Members of this vintage become caught up in the revolution, learning new truths that give them a knowledge edge. Such was the case for scientists trained during the early 1950s when the importance of nucleic acids as the fundamental biological molecule was just being discovered. Other generations have the bad fortune to be educated prior to a major change or breakthrough. Members of such a generation belong to a poor vintage and, unless they are able to switch gears and learn new ways of thinking, face knowledge obsolescence. Such was the case for scientists trained in the late 1930s and early 1940s to believe that proteins were the key to genetic information.

WHY THESE ISSUES ARE OF CONCERN

It is part of the human condition to wonder how much control individuals have over the course of events that shape their lives, how much their fate is determined by things outside their control. Thus, from the point of view of the individual scientist, it is important to know whether age is a fever chill: whether scientists really are washed up at 30 or 35.

Scientists are also interested in knowing the extent to which their success or lack of success is due to RPRT. Just how important is the scientist's vintage? Do certain vintages, educated at the right time (and at the right place) have a research edge throughout their careers? Are job market conditions really important? Does job location depend largely on what is happening in the market for scientists at the time their training is completed?

For nonscientists also, the issues of age and RPRT are important, since in recent years the American scientific community has aged considerably. One indication of this is the fact that from 1973 to 1987 the median age of Ph.D. scientists rose from 41 to over 44.[6] This may seem to be an insignificant change until one realizes that for most of the twentieth century, although individual scientists aged, the scientific community became progressively younger as more people were recruited into science than retired. By the early 1970s, however, this process had reversed itself, and the scientific community, like its individual members, began to age. The most revealing way to see this is to examine the number of scientists under the age of 35. In 1973 the figure stood at approximately one in four; today it has fallen to about one in ten.[7] In academe, where most basic research occurs in the United States, the change has been even greater. Thus, it is not surprising that the Nobel laureate Leon Lederman, when asked on the eve of 1990 to evaluate U.S. graduate education, spoke of "symptoms of deep trouble," largely related, the white-haired physicist observed, to the fact that in academe "everybody looks like me."[8]

It is not just science that is in trouble; the productivity of the United States also is at stake. Science affects the quality of our lives and certainly our competitive position in the world, and it affects our standard of living. Imagine yourself in a "Noyceless" world. Is an older scientific community holding back the United States? Has the aging of the scientific community contributed to the decline in the United States' productivity growth since 1973? Has the problem been exacerbated by job market conditions that, as we shall see, have left a disproportionate number of young scientists at farm teams? Has it been made even worse by changes in recent years that have led many of the best and the brightest to choose careers outside science?

HOW THE AMERICAN SCIENTIFIC COMMUNITY CAME TO BE OLDER

How did the United States arrive at a situation in which the scientific community aged significantly over such a short period of time? How did we create a situation in which a whole generation of scientists found it difficult to obtain jobs in the top research sector? Why are many of the most talented youth attracted into other fields, such as law and business? Although there is clearly more than one explanation, much can be attributed to public policies followed during the past forty years.

The success of the Manhattan Project and the GI bill gave science a tremendous boost after World War II. Money was available for education and for basic research. Science grew, and consequently the scientific community became younger. This process was enhanced tremendously by the United States' response in 1957 to the launching of *Sputnik* and the initiation of the space program. Almost immediately, the government began a massive program of support for science. Much of this support continued through the 1960s. During this period science grew at a fast and furious pace. In response, many states established new universities and colleges and expanded existing programs. Then, in the late 1960s, with the acceleration of the war in Vietnam, science budgets (along with others) began to be squeezed. First, money for research in the physical and earth sciences failed to keep pace with inflation. Then, beginning in the early 1970s, the government sharply cut graduate student stipends in the sciences. Those most affected were the engineering and physical sciences, in which fellowships and traineeships were cut by 90 percent.[9] As a consequence, jobs in science began to dry up. Universities retrenched and research institutions were slow to hire because they were "tenured up" and had fewer federal dollars to support research. As a result, more Ph.D.'s headed for jobs in business and industry. Ironically, at approximately the same time, the research and development effort in business and industry, which had grown at an impressive real rate of 6.4 percent in the 1960s, stalled.[10] Not surprisingly,

Introduction 7

promising students began to look elsewhere, often choosing careers in business, law, or medicine.

WHAT THIS BOOK IS NOT ABOUT

Before laying out the plan of this book, we take a moment to let the reader know what this book is not about. First, what we have to say is more pertinent to the American scientific community than to other communities. Although in many instances our arguments can be and are extended to scientists working in Western Europe, we make no effort to generalize to scientists working in other societies. Second, with rare exceptions, the evidence presented in this book concerns male scientists, largely because science, particularly the physical and earth sciences, has been dominated by men. Furthermore, the very fact that there are so few female scientists raised concerns about our ability to keep data for women scientists confidential had they been included in our own research. Third, much of our discussion is about scientists in academic settings. The reasons for this are that academic scientists have generally been at the forefront of basic research and that academic scientists typically publish their findings and thus leave a measurable research trail. Fourth, our discussion is restricted to the physical, earth, and life sciences, with occasional references to mathematics and engineering. We have made no attempt to study the social sciences. Finally, we should point out that most of the empirical work summarized in this book concerns Ph.D. scientists. Non-Ph.D.'s were not included in our work and also are not, with few exceptions, included in the studies done by others.

THE PLAN OF THIS BOOK

Chapter 2 sets the stage for what is to come, by considering how science is done and why scientists do science. This chapter stresses the importance of colleagues and equipment to scientific success. It also examines the enormous time commitment that science requires. Why scientists are willing to spend so much time doing science is also discussed. We offer three reasons. Scientists do science because it is a kind of puzzle, the solution to which is itself a reward. They seek to uncover the fundamental laws that govern how the universe operates. It is this interest in solving the puzzle, in obtaining the truth, that initially attracts many persons into science. Scientists also are interested in the glory that accompanies discovery. They crave recognition. They play a game in which being first, and the recognition given to being first, is extremely important. The ribbons of success are not the only extrinsic rewards, however, that motivate scientists. Scientists are also interested in the gold that is often attached to winning. In this respect, scientists are not unlike other people. They like money.

Chapter 3 looks at why there may be a right age for doing science. Drawing on the literature in sociology, psychology, and economics, we examine theoretical reasons for an age–research relationship. Of particular concern is whether the motivators of science (puzzle, ribbon, and gold) lead scientists to spend less time doing science as they age. We also consider the cognitive changes related to the physiology of aging and the relationship between age and creativity, focusing on why in certain fields creative work may be the domain of the young.

In Chapter 4 we turn from theory to the evidence concerning whether there really is a right age for doing science. Two issues are addressed. First, is there a relationship between age and the production of outstanding work by eminent scientists? Second, is there a relationship between age and productivity across the wider spectrum of Ph.D. scientists? Much of the evidence pertaining to the second question comes from a study that we conducted in the 1980s. This chapter also discusses the methodological problems that arise in estimating an age–productivity relationship, particularly whether aging effects can be disentangled from the effects of being part of a particular generation. For example, if the most recently educated are the best educated (in the sense that their knowledge is more up-to-date) and if younger scientists are found to be more productive, can we determine whether the younger scientists are more productive because of their age or because they come from a superior vintage?[11]

The focus of both Chapters 3 and 4 is the individual scientist. Chapter 5, on the other hand, looks at the consequences of an aging scientific community for the larger society. Here we examine whether an older community slows the speed with which new ideas are integrated into scientific theory and practice as well as whether an older scientific community retards the rate at which the economy grows. As we shall see, an older community may be less productive because its older members produce less science. An older community may also indirectly slow the progress of science by resisting the innovative ideas of the young. The consequences are felt not only by science but by the overall economy as well.

Chapter 6 considers the importance of RPRT for the individual scientist. If resources and colleagues are essential to scientific discovery, as Chapter 2 argues, do scientists of equal quality have equal access to these resources? Chapter 6 contends that they do not. Certain scientists have the good fortune to enter science at the right time and subsequently get to the right place for doing research. The chapter also expands the concept of vintage that we introduced. Just as some scientists have the good fortune to complete their training when jobs in the research sector are plentiful, some scientists also have an intellectual advantage, having been trained when intellectual conditions were favorable in their field. Other scientists are not as fortunate and must play catch-up if their knowledge is not to become obsolete.

Many of the conditions that lead to RPRT are not specific to the individual but, rather, to a generation. Thus we can speak of the generational consequences of RPRT. These generational consequences lead us to believe that certain "models" of scientists of otherwise equal quality may have more science

Introduction

in them than other "models" do. That is, certain models of scientists may be better than other models. It is not only because of RPRT, however, that certain generations are especially productive and others are not. Chapter 7 argues that the quality of persons choosing careers in science varies across generations. Thus, in addition to market and vintage effects, there are *quality effects*.

Chapter 8 presents our case study of the relationship among age, cohort ("model"), and scientific productivity. For this study we assembled a unique and rich database by matching records of Ph.D. scientists in the United States—taken from the *Survey of Earned Doctorates* conducted every other year by the National Research Council—with data on publishing activity taken from the *Science Citation Index* prepared by the Institute of Scientific Information.[12] The analysis is limited to scientists working in "top" academic research institutions employed in atomic physics, particle physics, solid-state/ condensed-matter physics, geology, geophysics, and oceanography. Although we have made every effort to be as nontechnical as possible, some readers may find the early material in this chapter slow going. Such readers are encouraged to skip ahead to the results section of the chapter.

Our major conclusions are presented in Chapter 9. Very briefly, these are the following:

1. Exceptional contributions in science, the type for which persons are awarded the Nobel Prize, are most likely to be made by scientists under the age of 40.
2. Age matters, but not nearly as much, for "average" Ph.D. scientists.
3. RPRT clearly matters, but unlike age, it is significantly more difficult to study empirically.
4. In the past twenty-five years, the average quality of Americans choosing careers in science has declined.
5. Job market conditions and quality considerations have conspired to make those who entered science in the late 1960s, 1970s, and 1980s less productive than their counterparts who became scientists at an earlier time.
6. Increased competition in U.S. science is acting to stymie creativity and productivity.

Given the importance of scientific research to economic growth, these conclusions imply that the overall productivity of the United States has been adversely affected by events that have taken place in science over the last twenty-five years. Moreover, because the lag between research and the effect on the economy is long—estimated to be from ten to thirty years—the piper has just begun to be paid.

The irony is that many of the conditions that have led to a decline in productivity have been caused by fluctuations in national policy. The final pages of Chapter 9, therefore, ask: Where do we go from here? Several avenues for change are discussed, three of which we note here. One is a commitment by the United States government to long-term real growth in federal funding of about 4 to 5 percent. Such a policy can be sustained and would eliminate the shock waves that a policy of +15 percent one year, −2 percent the next year, sends through the scientific community. Second, we believe that the United States would benefit if more of the fun were put back into science. Science has sometimes been characterized as play behavior carried into adulthood. In recent years, however, science has increasingly become more work and less play, and this has taken a toll. Third, we believe that the United States must rethink the way in which research, particularly university-based research, is organized and conducted. This means reassessing the role of graduate students in the research process and also reconsidering the role of federal funding. A case can be made that the mad chase for grants, and the long hours that it requires, has weakened the scientific enterprise.

Change in the way that science is done will necessitate a reallocation of resources, and reallocation is often a painful process. The pain, however, can be lessened if the United States takes advantage of the window of opportunity provided in the next fifteen years by the retirement of a large number of scientists. The age structure of the scientific community in the United States has worked against science for many years, but in the 1990s it can work to its advantage.

NOTES

1. Jack Kilby, an engineer at Texas Instruments, actually invented the integrated circuit before Noyce did. But Noyce's integrated circuit—made of silicon and not germanium, as was Kilby's—set the standard for the industry. Thus Noyce became known as the coinventor of the integrated circuit (see Wolfe 1983).

2. Weber 1980, p. 162.

3. Walter Brattain joined Bell Laboratories in 1929 at the age of 27. The invention of the transistor in 1947 was the culmination of a long research program begun at Bell in 1931-2 and interrupted by World War II. Early discoveries included the finding that silicon was a good semiconductor. Brattain's coinventors were John Bardeen, who joined Bell in 1945 at the age of 37, and William Shockley, who joined Bell in 1936 at the age of 26.

4. Zuckerman 1977, p. 164.

5. Rigden 1987, chap. 1.

6. Special SDR tabulations for U.S.-trained doctoral scientists (excluding engineering, psychology, and the social sciences) employed full time in science and engineering in the United States.

7. Survey of Doctorate Recipients, various dates.

8. Interviewed on the "MacNeil/Lehrer News Hour," December 26, 1989.

9. U.S. Congress 1989.

10. Scherer 1986, p. 19.

11. This assumes that these scientists are observed at the same point in time.

12. Details concerning this database can be found in Stephan and Levin 1987.

2
How Science Is Done; Why Science Is Done

Our goal in this book is to investigate the importance of age, time, and place to scientific productivity. In order to do so, we need a framework for analysis because, as we shall see in the chapters to come, the productivity consequences of age, place, and time are indirect, just as they are in other situations. For example, if we want to know why older persons get cataracts or have gray hair, we must first know something about biology. And if we want to know why 1949 was an exceptional year for Bordeaux wine, we must know something about growing grapes. Likewise, for our purposes we need a basic understanding of how science is done and why scientists do science.

HOW SCIENCE IS DONE

At the core of science are scientists. Generally, however, scientists do not work alone, nor do they work without some tools of the trade. These vary from paper and pencil to a Cray supercomputer to an accelerator that moves particles to energies of 800 GeV (GeV represents "giga" or billions of electron volts). Economists modeling the scientific process think of these components as inputs into the production of science, in which the quantity and quality of output produced depend on both the inputs and the proportions used. In our analysis of how science is done, we discuss each of these components in some detail. We begin with the inputs that scientists bring to the process.

The Scientists

Scientists contribute two dimensions to the scientific output: effort and cognitive resources. One dimension of effort is time. Although it is popular to characterize scientists as having instant insight, science in fact takes time. No one knows just how long, or how many times, Archimedes sat thinking in his bathtub before shouting "Eureka!" "Any scientist who is not willing to put in the hours formerly reserved for factory workers in Victorian England is not likely to succeed."[1] Anne Roe, in her interviews with scientists in the early 1950s, found that many of her eminent subjects worked nights, Sundays, and holidays, and quite a few did not take regular vacations.[2] The importance of time in scientific discovery is often expressed by scientists and reported in the chronicles of breakthroughs in science. Stanley Cohen, a Nobel Prize–winning biochemist, for example, is reported to have remarked, "If I manage to solve a problem it's only because I've really plugged away at it. I have to work hard, very hard, to find the solution."[3] James Gleick, the chronicler of the emerging science of chaos, reports that when Mitchell Feigenbaum began to work on the concept of universality, he worked twenty-two hours a day for two months, existing almost exclusively on coffee.[4] And Robert Hazen, who relates the story of the race for high-temperature superconductors, tells of how, when the competition was at its zenith, scientists literally worked around the clock in their laboratories.[5]

There is another essential dimension to effort, and that is motivation, the dedication to do science. Mary Frank Fox, a sociologist of science, notes that certain investigations have shown that "productive scientists, and eminent scientists especially, are a strongly motivated group of researchers" and have the "'stamina' or the capacity to work hard and persist in the pursuit of long-range goals." She goes on to say that "informed observers have long described high-producing scientists as driving and indefatigable workers. And empirical data on scientists and engineers do suggest that high performers are absorbed, involved, and strongly identified with their work."[6]

Scientists not only bring effort to science, they bring cognitive resources as well. One dimension of these cognitive resources is the scientists' knowledge base. Are the scientists familiar with Julia sets,[7] work in crystallography, field theory? Do they understand renormalization? Are they equipped to use the latest in laser technology, a scanning–tunneling microscope? A related and perhaps more significant dimension is whether the scientists' cognitive resources facilitate learning new ideas as they emerge. Have they been trained in the kind of knowledge that allows easy access to other knowledge? Can they "cross over" to new areas, perhaps outside their initial domain, as they evolve?

The other side of cognitive resources is ability, in particular the ability to do science. Scientists on the whole are smart. Lindsey Harmon reports that Ph.D. physicists have an average IQ in the neighborhood of 140.[8] Catherine Cox, using biographical techniques to estimate the intelligence of eminent scientists, reports IQ guesstimates of 205 for Leibnitz, 185 for Galileo, and 175 for Kepler.[9] It may be precisely because scientists are so smart that the two studies that have looked for a relationship between IQ and scientific performance have failed to find one that is statistically significant.[10] Variation in IQ among those already at the top of the intelligence distribution may explain little of the variation in performance. The lack of a statistical relationship may also indicate that standard IQ tests do a poor job of discerning the ability to do science. Roe found that the scientists she interviewed were very smart, but she also discovered a wide range in the group in performance on standard intelligence tests. Yet all the men in her sample were judged eminent at the time of the study.[11]

Although the evidence concerning a relationship between IQ and scientific productivity is sparse, there is a general consensus that certain people are particularly good at doing science and that some are not just good but superb. They have a sort of creativity that seems "to flow from some magic gland."[12] There is a belief in science not only that some have this magic gland but also that those in the selection process can sometimes identify persons with it *ex ante*, as opposed to *ex post* when everyone knows. Two examples readily come to mind: James Watson and Kenneth Wilson, both Nobel laureates. While Watson was still a student at Indiana University, his professor, Salvador Luria, himself a future laureate, reportedly introduced Watson, then not yet 20 years old, as the most brilliant of his students.[13] As for Kenneth Wilson, at the time he came up for tenure at Cornell University he had virtually no published work to his credit. Yet enough of the Cornell faculty perceived him as having a "deep capacity of seeing into physics" to award him tenure anyway.[14] Soon after, the papers began to pour out, one of which was to win for Wilson a Nobel Prize in 1982. On a broader scale, this belief in the magic gland is what makes prestigious academic departments persistently search for extraordinarily talented young scientists.

The Importance of Equipment

Like science in the eighteenth and nineteenth centuries, modern science is heavily influenced by the availability of technology. In the world of science, the bird who gets the worm is not necessarily the brightest bird or the hardest-working bird, but the bird with the best equipment for worm spotting and worm digging. The history of science is a history of just how important resources and equipment are to discovery. If this is true historically, it is, with few exceptions, even truer today. According to the philosopher of science Ronald Giere, "The overwhelming presence of machines and instrumentation must be one of the most salient features of the modern scientific laboratory. . . . The development of science depends at least as much on new machines as it does on new ideas."[15] In one of his last public lectures the historian of science Derek de Solla Price stated,

> If you did not know about the technological opportunities that created the new science, you would understandably think that it all happened by people putting on some sort of new thinking cap. . . . The changes of paradigm that accompany great and revolutionary changes may sometimes be caused by inspired thought, but much more commonly they seem due to the application of technology to science.[16]

This definitely is the case with the proton, which, at the beginning of this century, was such an abstract concept that "scientists had real questions about the reality of any such thing." Now "the proton has been tamed and harnessed to the equipment used to investigate other particles and structures: quarks, gluons, and the shell model of the nucleus."[17] On a much smaller scale, the diamond anvil developed over the past ten years by geologist Jim Stout and a colleague at the University of Minnesota incorporates materials that were not available to scientists until the mid-1970s and opens up the possibility of finally studying rocks located seven hundred miles below the earth's surface.[18]

It is not only experimental scientists who use equipment. Despite the view that the purest scientists do not need equipment—having sufficient equipment in their heads—theorists have become increasingly dependent on very large-scale computers. This is true because much of the theorists' work comes from modeling mathematical systems for which the required calculations are generally much too complex to complete by hand or to have been completed by the types of computers available in the 1950s, 1960s, and early 1970s. A great deal of theoretical work thus requires access to very expensive machines.[19]

The importance of equipment in science is reported again and again in accounts of scientific discovery. Galileo had his telescope; Boyle used an air pump; James Watson first saw the B pattern for DNA through X-ray diffraction pictures taken by Rosalind Franklin. When Watson and Crick thought they had the double helix, they ordered the lab to build the components that would allow them to construct a three-dimensional model. Measurement of the magnetic field along the ocean floor, which provided essential data for the emerging theory of plate tectonics, was greatly facilitated by the development of the proton precession magnetometer.[20] Perhaps nowhere is the role of equipment more obvious than in experimental particle physics, in which the introduction of accelerators operating at higher and higher levels of energy has opened up an inward world that scientists only dreamed of a few years ago.

Some of the equipment used in science carries bargain-basement prices: Gregor Mendel used peas; T. H. Morgan used fruit flies; early researchers in the science of chaos used Apple computers. But most of the equipment, unfortunately, does not carry bargain-basement prices. Indeed, even the "tabletop" experiments done in acoustics use equipment that costs in excess of $100,000, and most labs, even small labs, require substantially more expensive equipment. It is not at all uncommon for a scientist, even in a low-expenditure area of science such as horticulture, to have a lab with a half-million dollars of equipment. And this is just the beginning, just the low end. At the high end of the spectrum are accelerators for detecting particles, the newest one of which—the Superconducting Supercollider—is being constructed after much fanfare with a price tag at this writing in excess of $8 billion and still rising. Somewhere in between are giant telescopes costing in the neighborhood of $85 million and Cray supercomputers with a price tag between $2.5 million and $23.7 million, depending on the number of processors.

Access to expensive equipment comes through institutions, first in graduate school when the student works in a professor's lab, then as a postdoctorate (generally at another institution), and finally at a business or institution where the scientist's own research agenda is activated. Some scientists attend graduate schools that are well endowed; some get jobs in institutions with good labs; some get grants to develop their own labs; and some have no trouble getting funded to spend time on an accelerator or at a major telescope. But others do not. One theme of this book is that the type of equipment to which scientists have access affects their productivity. A related theme is that not all scientists have equal access to equipment, as not all scientists have equal success in getting jobs at places that are either flush with resources or flush with the prestige that can be leveraged into resources.[21]

The importance of equipment in scientific discovery also relates, as we shall see, to the obsession of scientists with recognition and priority. Once the equipment exists that is required to understand phenomenon X—to pick the apple from the tree—then others can pick the apple as well, and so the scientist has enormous incentive to get there first (or plant a private orchard) and let the world know.

The Importance of Colleagues

Good science is often done in communities of scientists and often in teams. Indeed, the growing importance of the team in research is one of the major trends in science in the twentieth century. Everywhere we look we see teams: teams trying to unravel the genetic code, teams working on superconductivity, teams in chaos research. Nowhere are teams more evident than at accelerators, where a "team" may have as many as five hundred players, and the author list may be longer than the article.

Teams have become increasingly important in science because of spiraling specialization and an enormous emphasis on equipment that requires even more specialization. In modern science, "many problems require an array of cognitive resources that no single scientist is liable to possess."[22] Accordingly, team members have access not only to their own cognitive resources but also to the cognitive resources of others. Consequently, the team may be able to produce more than the individual parts. The literature on scientific productivity points to the idea that scientists on teams are more productive, oftentimes producing "better" science than lone players do.[23]

The power of the team was described by Rita Levi-Montalcini. In trying to identify the "nerve-growth promoting agent," Levi-Montalcini found that her lack of training in biochemical techniques stood as an impediment. Then she met Stan Cohen, a biochemist. About their collaboration, which eventually won them a Nobel Prize, Levi-Montalcini says: "The complementarity of our competences gave us good reason to rejoice instead of causing us inferiority complexes." She recalls that Cohen commented, "Rita, you and I are good, but together we are wonderful."[24]

Other well-known collaborators include François Jacob and Jacques Monod, and Francis Crick and Sydney Brenner. Jacob and Monod won the Nobel Prize in medicine in 1965 and had the habit of speaking to each other almost daily for two to three hours.[25] Crick and Brenner shared an office for twenty years and had the rule of uttering "anything that came into your head." Although most of these conversations were, according to Brenner, "just complete nonsense . . . every now and then a half-formed idea could be taken up by the other and really refined."[26]

Teams do more for each other than share cognitive resources; they also share psychological resources. Team members reinforce one another's drive to achieve. The superconductivity players discussed whether their team would be first. Team membership can also benefit shy scientists who might otherwise not stand up for their ideas in the competitive world of science in which "selling" is part of the game. According to Gleick, the young physicist Robert Stetson Shaw, a key player in chaos research, "suffered from a certain diffidence in putting his ideas forward in the academic marketplace; fortunately for him, his new associates (members of the self-named Dynamical Systems Collective) had no such problem."[27] Teams, and more loosely structured research groups, also increase the visibility of one another's work, minimizing what the philosopher of science David Hull calls the "greatest danger for any bright idea," the danger that the idea will be ignored.[28]

Research groups are composed of more than peers. Most teams include graduate students and postdoctorates. Indeed, the typical pattern for scientists in the United States is to work first as a graduate student and then as a postdoctorate in a professor's lab on a project designed by the professor under a grant written by the professor and to publish joint papers with the professor and other collaborators. This points to another skill that the leader of a team must possess: the ability to write successful research proposals that will support graduate students and postdoctorates and also purchase new equipment. For example, a scientist doing tabletop research with five or six graduate students would find that such a "modest" research agenda would require funding at the level of $250,000 plus. Writing proposals takes both skill and time. For example, it takes about one hundred hours just to complete a National Institutes of Health grant application, and many scientists write more than one proposal a year.

Cooperation in science extends far beyond team players, applying also to solo players. Regardless of whether they are team members, scientists talk with other scientists, sharing ideas, critiquing one another's work, and cheering on the others to victory. This occurs in informal ways over coffee or lunch and in formal presentations of seminars and papers. The interchanges that result from such discussions can make spectacular differences in science. Remembering how his office mate, Jerry Donohue, kept him from taking a faulty route, Watson recalls, "If he had not been with us in Cambridge, I might still have been pumping for a like-with-like structure."[29] Giere, in describing Fred Vine's considerable contribution to plate tectonics, states that "what made the difference for Vine, apart from his considerable cognitive resources, was the additional material resource of being in a place [Cambridge] that attracted people like [Harry] Hess and [J. Tuzo] Wilson."[30]

All of this means that place does matter. Scientists working alone and with few resources can be productive—perhaps the best example is Mendel and his path-breaking research in genetics—but the examples are few and far between. To do science requires individual cognitive resources and access to equipment and also access to the intellectual and psychological resources of others. Clearly these resources are not randomly scattered across a nation but are heavily concentrated in major research centers. One of the persistent findings in the sociology of science is that scientists in "prestigious" academic departments produce the lion's share of the research. Studies imply that this happens not just because these institutions have an eye for good talent (which they do) but also because these institutions provide an environment favorable to research, which, in addition to equipment, includes stimulating colleagues and good graduate students.[31]

The importance of colleagues in science does not stop at the boundary of the particular institution with which the scientist is affiliated. The productive scientist is also likely to belong to an "invisible college," a group of approximately one hundred scientists who share common interests, exchange preprints, and meet formally at conferences and informally, often in pleasant places such as Aspen, to exchange ideas. These invisible colleges — christened as such in this century by Price after a concept posited by Boyle in the seventeenth century — play a significant role in science by furthering knowledge, establishing research agendas, and adjudicating issues of priority.[32] It was at such a "college" that Watson first met Maurice Wilkins. Through such a college network, much of the research on chaos has spread.

The Role of Serendipity

Finally, in talking about the ingredients needed to do science, we also must acknowledge the role of chance, and particularly the role of the "happy accident," serendipity.[33] Serendipity clearly has a place in science when unanticipated events occur and scientists, following up on why they occur, make new discoveries. The following up, of course, is not an accident. "Chance," according to Louis Pasteur, "favors only the prepared mind."[34] Edward Lorenz, a pioneer in the field of chaos, took a shortcut in his research that serendipitously opened up a whole new avenue of investigation.[35] Equally serendipitous for Sir Alexander Fleming was the wind blowing in the penicillin mold.[36] Other examples of chance discovery include X-rays, insulin, nuclear fission, electric current, and synthetic rayon.

WHY SCIENCE IS DONE

Joe Ford, a chaos researcher at the Georgia Institute of Technology, is reported to have observed: "To me chaos is like a dream. It offers the possibility that, if you come over and play this game, you can strike the mother lode."[37] Ford's statement is a good articulation of the reasons that scientists give for doing science: to solve the puzzle and to win the game. The distinction between the two is that puzzle solving involves a fascination with the research process itself, while winning the game offers recognition among fellow scientists. The successful scientist hopes to achieve intrinsic rewards while doing research and extrinsic rewards after completing it. Both motivating factors play heavily on the metaphor of gaming, and both imply that there is an attainable solution to the question being investigated.

The Puzzle-solving Aspect

The importance of the research process itself as a motivating force is discussed extensively by Thomas Kuhn[38] and summarized well by the sociologist of science Warren Hagstrom: "Research is in many ways a kind of game, a puzzle-solving operation in which the solution of the puzzle is its own reward."[39] The particle physicist and Nobel laureate Richard Feynman speaks of the "puzzle drive,"[40] and a friend described Feynman's "almost compulsive need to solve puzzles."[41] A theoretical physicist told Hagstrom, "[I get more pleasure] when the problem has been solved in principle but when hard work remains to be done—when you have enough security to know you're not wasting your time but while there is some challenge left in the problem."[42] An experimental physicist reported that you receive the most gratification "when you find the effect you're looking for—everything else is anticlimax."[43] Solomon Snyder, who, along with Candace Pert, identified the opiate receptor, speaks of research as a joy; Hazen talks of the fun that the researchers had in their quest for high-temperature superconductors.[44] Hull speaks of the innate curiosity of scientists, noting that science is "play behavior carried to adulthood."[45] He continues: "The wow-feeling of discovery, whether it turns out to be veridical or not, is exhilarating. Like orgasm, it is something anyone who has experienced it wants to experience again—as often as possible.[46]

For some scientists, the hunt for answers provides sufficient incentive to engage in research. Examples include Mendel, the reclusive monk whose path-breaking work on the genetics of pea plants was published in 1866 in an obscure journal and went unnoticed until 1900, sixteen years after his death, and Srinivasa Ramanujan, an uneducated Indian holding a minor clerical job, who over a ten-year period produced numerous elegant insights to existing mathematical theorems and formulated and proved new theorems, without any contact with the community of mathematicians.[47] Certainly the history of the science of chaos is a history of scientists motivated to work in the area, at least initially, because of their desire to solve a puzzle.

Recognition

The psychological rewards from doing science, however, are probably not sufficient to motivate most scientists. Scientists crave recognition, and in this they are no different from their fellow human beings. "The pursuit of reputation in the eyes of others is," according to Rom Harré, "the overriding preoccupation of human life."[48] "Give me enough ribbon," Napoleon is reported to have said, "and I can conquer the world."[49]

Robert Merton and the sociologists and philosophers of science who have followed him have demonstrated that recognition is one of the main factors leading scientists to do research and have emphasized how this taste for recognition, if not acquired in graduate school, is at least nurtured through the socialization process when one is a student.[50] Moreover, scientists do not crave recognition of the Johnny Carson or *Newsweek* variety. Indeed, such mass recognition can even, it is argued, jeopardize a scientist's standing among his or her peers. David Raup[51] coined the phrase "saganization" to describe the loss of professional reputation that a scientist (such as Carl Sagan) suffers after receiving continued mass media attention. Rather, what scientists want is recognition from their peers in the form of citations of their work, invitations to speak at important gatherings, appointments to prestigious departments, and awards. Charles Darwin once said, "My love of natural science . . . has been much aided by the ambition to be esteemed by my fellow naturalists."[52]

Clues to the importance of reputation as a motivating force in science are apparent everywhere. The order of authors on articles is not haphazardly arrived at but is something that is often carefully negotiated.[53] Hazen describes the care with which the author order was established for articles being written on superconductivity.[54] Scientists read footnotes to see whether they were thanked in someone else's work. Snyder reveals that soon after the publication of the Pert-Snyder paper, Avram Goldstein wrote him noting that "if I had any criticism, it would be that you were a little ungenerous about the scientific and intellectual precedents of your work."[55]

Nowhere is the importance of reputation to scientists more apparent than in issues concerning priority of discovery. Science is not just a game played for one's enjoyment. It is a game played to win. And as Jerry Gaston has pointed out, unlike many other competitions, it does not award second and third prizes.[56] As a result, scientists are obsessed with priority, with establishing that they reached the pinnacle first.[57] For science, claims are staked not by flags on top of peaks but by articles rushed into print before others can get there. "Rushed" in the literal sense.[58] It is not unknown for articles to be written and posted on the same day. Hazen, describing the race for superconductivity, tells how scientists at Bellcore had someone drive their manuscript to the journal's office rather than put it in the overnight mail, thereby ensuring that they would be one more step ahead in the battle for priority.[59] And after a scientist has had an article accepted in the prestigious journal, it is common to negotiate with the editor for a "note added" so that work completed between the time of submission and publication can be reported, making the claim to priority all the more convincing. Concern with priority also leads scientists to communicate results via electronic mail and to present findings at meetings long before they appear in print. And at meetings it is not uncommon for the speakers to compete for placement on the program. "People have become hysterical over when in the order of speakers they come. . . . It is best to be scheduled at the beginning of a meeting so that if others have done the same experiment as you, you will have reported it first and they will have to refer to your work."[60] The significance of priority can also cause scientists to go public "too soon," before results can be verified. Such appears to have been the case for the Utah cold fusion team in 1989.

The extreme importance of not just solving the puzzle but also reaching the pinnacle first—and the consequences that it has for motivating scientists to do science—is spelled out clearly in Watson's account of Crick's and his effort to beat Linus Pauling in the race for the DNA structure. In this climb, one team reached the summit well ahead of the other. Sometimes, however, scientists get to the summit almost simultaneously, and then the issue of who got there first, of establishing priority, becomes all-consuming. Sometimes the debate over priority goes public, as in the case of the dispute between Luc Montagnier of the Pasteur Institute and Robert Gallo of the National Cancer Institute concerning the discovery of the AIDS virus. Oftentimes, however, priority disputes are a much more private affair, taking place behind the closed doors of the scientific community. These debates can be particularly difficult to referee when researchers have access to one another's work prior to publication, as in the case of the discovery of insulin[61] or the work on plate tectonics in which Fred Vine was sent data on the *Eltanin*-19 profile (magnetic data of the ocean floor obtained by the research vessel *Eltanin*) without the consent of Walter Pitman, and then Vine used the data to prepare a paper. Pitman, not wanting to be scooped, suggested to Vine that they coauthor the paper, a suggestion that Vine rejected. Eventually the editors of *Science* were brought into the debate. Although the editors wished to publish both papers simultaneously, they agreed to publish Pitman's two weeks before Vine's, prompting Pitman to remark that "the guys that run *Science* understand geopolitics."[62]

A different side of scientists' concern for reputation is what Hull calls the desire to show that "son of a bitch." In his book, Hull recounts how the scientists he interviewed were spurred into feverish activity to ensure not only that they won the race but that someone else did not. Victory is sweeter if there is not only a winner but also a loser to be pointed out. This is particularly true when a scientist has previously lost a race to a specific individual or the scientist's professional reputation has been called into question by another scientist. Finally, lest one think that priority debates are a new phenomenon in science, one should realize that they go back for centuries. Merton details numerous cases occurring over the past three hundred years.[63] Indeed, it has been argued that disputes over priority are declining in science. This should not, however, be taken as an indication that scientists have become less interested in recognition but, rather, that they, through modern technology, are more capable of planting their own orchards so that they can be sure that only they will pick the apples.

Monetary Rewards

Puzzle solving and an interest in recognition are strong motivating forces for doing science. They are not, however, the only motivating forces. In addition to ribbon, the tangible rewards for doing science include gold. And scientists, despite what the public may want to think, are indeed motivated by money. When Henry Rosovsky, upon becoming dean of the Faculty of Arts and Sciences at Harvard, asked one of Harvard's most eminent scientists the source of his scientific inspiration, the reply (which "came without the slightest hesitation") was "money and flattery."[64]

Scientific productivity is financially rewarded by institutions outside the employing relationship as well as by the employer. One significant study of the academic world in the 1970s concluded that "whether or not the 'publish or perish' mechanism is operating, 'publish and flourish' clearly does. There is no question that publication sharply enhances academic men's chances of high salary — and also of earnings outside the university."[65] External rewards come in the form of prize money, royalties, and consulting and speaking fees. In addition to recognition, awards in science can be monetary, the most obvious being the Nobel Prize, with a 1990 purse of about one million dollars. Answers to the puzzle can also have substantial economic value, as scientists may stake their claim through the patent process, thereby opening up the possibility for collecting substantial patent royalties. Being highly productive also creates opportunities to speak and consult — to go on the "circuit" — and collect the fees attached to such services. Snyder, for example, found that after he and Pert identified the opiate receptor, a large number of invitations to lecture at universities and at scientific meetings came his way.[66] Oliver Fulton and Martin Trow report that highly published academics are more likely to be consulted than are academics who are not highly published.[67]

Productive scientists also earn more "at home." There is substantial evidence that salary is directly as well as indirectly related to productivity. Howard Tuckman, for example, found that publishing articles brings significant salary increments in academe in engineering, mathematics, and physics. In chemistry and earth science, article publication is related to salary for those who have published more than fifty articles.[68] Arthur Diamond found salary to be positively related to career publications in his study of mathematicians employed at Berkeley between 1965 and 1977.[69] The financial consequences of this are impressive, for when salary adjustments are made for publishing performance, they are generally built into the salary base and therefore accumulate over time, not just in the year received. Putting the Diamond figures in 1991 dollars, his findings imply that for a 35-year-old mathematician, the present value of publishing another article is about $6,000.[70] Another indication that employing institutions reward productivity through higher salaries is found in the substantial amount of evidence that salary is related to the number of citations of an author's work. Diamond, for example, estimates that to a physicist an additional citation is worth about $100 (in 1991 dollars), or assuming that this increment is added to the base salary, for a 35-year-old, receiving one more citation has a present value of about $2,000.

Salary is related to productivity in indirect ways as well. For example, more productive scientists are more likely to be promoted, and promotion generally carries its own financial reward. The productive scientist may also enjoy a substantial salary premium, because productivity makes the scientist attractive to other institutions; hence salary reflects not only what the employing institution perceives to be the worth of the scientist but also what other institutions are willing to pay to attract the scientist away.[71]

One could argue that these monetary rewards are incidental and do not serve as a motivating force to do science, that scientists work for ribbon and love of the puzzle. But the evidence, although somewhat indirect, belies this conclusion. Through their actions, scientists indicate an interest in monetary rewards. For example, Richard Freeman found that the number of people going to graduate school in science accords with the salary situation in science.[72] Beverly Porter reports that 30 percent of the 1973 postdocs in physics stated five years later that they would not go into physics if they had the choice again. Many wished they had gone into medicine, presumably because of its higher pay.[73] The rush for high-temperature superconductors was and continues to be intense not just because of recognition but also because of money. The team that wins will be not only famous, but rich and famous, and the teams in the competition know this to be the case. Robert Noyce acknowledged "greed" to be a major motivator in his cofounding of Fairchild Semiconductor. For him, a driving force was "the prospect of a phenomenal return—if you won, you got several years' salary at once."[74] Surely the interest in cold fusion is not just for recognition but also for the enormous amount of profit in the offing. Finally, anyone who has ever watched the fights that can ensue in a university concerning the issue of raises and the reward structure cannot but believe that scientists want money as well as recognition.

In concluding this discussion of the "why" of science, we should note that the view of scientific motivation presented here is not a view dear to the public. Indeed, in a recent review of a book about Erwin Schrödinger, the physicist who discovered wave mechanics in the 1920s, Dick Teresi talks of how extraordinarily "common" Schrödinger was, implying that this commonness is unusual for a scientist: "The picture that emerges . . . is one of a conceited, selfish, childish, hopelessly middle-class nerd, one who worried about his awards and medals and was obsessed with his pension and salary. . . . He even drove a BMW."[75] Contrary to what Teresi and others may think, scientists, particularly productive scientists, are indeed common: They are concerned with reputation and even money. They want, as Stephen Jay Gould points out, "status, wealth and power, like everyone else."[76] They can exhibit bias, jealousy, and irrationality. Indeed, there is evidence that one can be "too nice" to do science; as an employee at the Stanford Linear Accelerator says, only the "blunt, bright bastards" make it.[77] To quote Hull, "the least productive scientists tend to behave the most admirably, while those who make the greatest contributions just as frequently behave the most deplorably. You pays your money; you gets your choice."[78]

NOTES

1. Hull 1988, p. 160.
2. Roe 1952, p. 58.
3. Levi-Montalcini 1988, p. 162.
4. Gleick 1987, p. 179. Chaos is the study of complex systems, such as the weather or heart rhythms, that become highly unpredictable over time. These systems are extremely sensitive to initial conditions and can turn "chaotic" when just one starting condition is altered slightly.

5. Hazen 1988. Superconductivity is the ability of certain materials at extremely low temperatures to conduct electricity with no resistance and thus no loss of energy. For practical applications, the quest has been to discover materials that are superconductors at higher and higher temperatures.
6. Fox 1983, p. 287.
7. Julia sets and their importance are discussed by Gleick 1987, pp. 216, 221-2, 228-9, and 236-7.
8. Harmon 1961, p. 679.
9. Roe 1952, p. 155, summarizes many of Cox's findings.
10. Bayer and Folger 1966; Cole and Cole 1973.
11. Roe 1952.
12. Gleick 1987, p. 160.
13. Levi-Montalcini 1988, p. 138.
14. Gleick 1987, p. 160.
15. Giere 1988, p. 138.
16. Price 1986, p. 247.
17. Giere 1988, p. 140.
18. See University of Minnesota, *Update*, February 1988, p. 4.
19. There is, however, an exception to the equipment rule. The science of chaos that has developed in the past thirty years has done so with little equipment, often only with what is in the scientists' heads. Indeed, although Gleick (1987) does mention the role of equipment in developing chaos (computerized images have been particularly important to understanding chaotic patterns), much of Gleick's book is devoted to showing how developments in chaos have come about by theorists working alone, without the aid of massive equipment.
20. Giere 1988, p. 248. Plate tectonics is the theory of a dynamic earth: that the earth's surface is constantly changing and that no feature on earth is permanent.
21. It is not always true that prestigious institutions provide excellent resources for their scientists. What they do provide, however, is exceptional access to the grants system, which, in turn, endows researchers.
22. Hull 1988, p. 514.
23. See Andrews 1979; Lawani 1986; Lindsey 1980; and Pravdić and Olvić-Vuković 1986 for a discussion of the relationship among quantity, quality, and collaboration in science.
24. Levi-Montalcini 1988, p. 163.
25. Wolpert and Richards 1988, p. 5.
26. Ibid., p. 106.
27. Gleick 1987, p. 248.
28. Hull 1988, p. 395.
29. Watson 1968, p. 133.
30. Giere 1988, p. 257.
31. See Blackburn, Behymer, and Hall 1978; Blau 1973; Long 1978; Long and McGinnis 1981; Pelz and Andrews 1976.
32. See Price 1986, chap. 3.
33. Chance plays a variety of roles in science. See Austin 1978; Barber and Fox 1962; Merton 1957, 1961; Turner and Chubin 1979.
34. Quoted in Turner and Chubin 1979, p. 440.
35. Lorenz found that the weather system that he was simulating on his computer displayed a totally unexpected pattern when, in order to save time, he rounded off to the thousandth place the values of the system's variables (Gleick 1987, p. 16).
36. Wolpert and Richards 1988, p. 6.
37. Gleick 1987, p. 306.
38. Kuhn 1970, pp. 35-42.
39. Hagstrom 1965, p. 16.
40. Feynman 1985, p. 9.
41. Ibid., intro.
42. Hagstrom 1965, p. 16.
43. Ibid.

44. Snyder 1989, p. 196; Hazen 1988, p. 248.
45. Hull 1988, p. 306.
46. Ibid., p. 305.
47. See Ziman 1976.
48. Harré 1979, p. 3.
49. Attributed to Napoleon by Menard 1971, p. 195.
50. See Merton 1957, for example.
51. Raup 1986, p. 164.
52. Merton 1957, p. 455.
53. Zuckerman 1968, however, found that the author order can be ambiguous.
54. Hazen 1988.
55. Snyder 1989, p. 69.
56. Gaston 1971.
57. In Watson and Crick's renowned paper on the structure of DNA, a *single* sentence was inserted concerning the genetic implications of the double helix to stake their claim to priority. "It has not escaped our notice that the specific pairing we have postulated immediately suggests a possible copying mechanism for the genetic material" (Crick 1988, p. 66).
58. For one discussion of the rush to publish in molecular biology, see Roberts 1991.
59. Hazen 1988, p. 69.
60. Gina Kolata, "Keeping the Scientific Show on the Road," *International Herald Tribune*, January 5, 1990.
61. Bliss 1982.
62. See Giere 1988, p. 274; Glen 1982, p. 337.
63. Merton 1957.
64. Rosovsky 1990, p. 242.
65. Fulton and Trow 1974, p. 61.
66. Snyder 1989, p. 83. Snyder chose to turn down the majority of the invitations, finding that the opportunity to become a jet-set scientist interfered with family and career.
67. Fulton and Trow 1974, p. 62.
68. Tuckman 1976, pp. 72–73.
69. Diamond 1986b.
70. The concept of present value is discussed in Chapter 3. These calculations assume a discount rate of 7 percent and an inflation rate of 4 percent over the remaining thirty years of the scientist's career.
71. Many of the indirect consequences of productivity on salary are reviewed by Tuckman and Leahey 1975.
72. Freeman 1975.
73. Porter 1979b, p. 153.
74. Wolfe 1983, p. 374.
75. Teresi 1990, p. 14.
76. Wolpert and Richards 1988, p. 146.
77. Traweek 1988, p. 146.
78. Hull 1988, p. 32.

3
Why Age May Matter

> A person who has not made his great contribution to science before the age of thirty will never do so.
>
> ALBERT EINSTEIN[1]

There is a great deal of anecdotal evidence that Albert Einstein was right, that science is the domain of the young. Charles Darwin was 22 when he set sail on the *Beagle*, 29 when he developed the concept of natural selection. Sir Isaac Newton was 24 when he began his work on universal gravitation, calculus, and the theory of colors. Karl Gauss was a mere 18 when he developed the method of least squares, and Sir Lawrence Bragg was a 22-year-old student at Cambridge when he developed X-ray crystallography. Fred Vine was 24 and a first-year graduate student at Cambridge when he formulated what was to become known as the Vine–Matthews hypothesis, a major step toward the articulation of plate tectonics.[2] Joshua Lederberg discovered sexual recombination in bacteria at the age of 21; Gerard 't Hooft was 25 when in 1971 he wrote an eleven-page paper that made a significant contribution to the renormalization of gauge theory and "ushered in the full panoply of the new physics."[3] James Watson was 25, and Francis Crick 36 (his career had been interrupted by World War II) when they uncovered the double helix. Einstein himself was 26 when he formulated the special theory of relativity, and P. A. M. Dirac, whose "fever chill" verse was quoted in Chapter 1, was 25 when he formulated his mathematical theory describing the relativistic electron. Other examples of youthful discovery include Werner Heisenberg, Wolfgang Pauli, and Niels Bohr. Indeed, so many of the early discoveries in quantum physics were made by the young that the field became known as *Knabenphysiks* (boy physics).[4] It has even been pointed out that Heisenberg and Dirac were of such a tender age when they were awarded the Nobel Prize that they were accompanied by their mothers to Stockholm to accept their awards. This seems a bit farfetched, however, as both were 31 at the time.

The idea that science is a young person's game is a view to which many scientists subscribe, although few take as extreme a position as Dirac and Einstein do. Other, less rabid believers include the Nobel laureate Abdus Salam, who stated in an interview on BBC that "there is a premium on youth" in physics,[5] and J. Robert Oppenheimer, who told Edward R. Murrow that in physics, great scientists have often done their most important work at early ages.[6] James Watson is also a believer. In an interview for the *New York Times* when he was in his forties, he stated that "almost every important discovery comes from someone under 35."[7] Robert Noyce agreed, insisting that on the electrical frontier great flashes of insight come to the young.[8] It is not only scientists who hold this perspective. The general public, at least the educated public, has subscribed to the idea ever since the psychologist Harvey Lehman published his massive and influential study *Age and Achievement* in the early 1950s.[9]

Anecdotal evidence and poems about fever chills do not, of course, a law make, and counterexamples exist. Schrödinger was 37 when he locked himself into a villa in Switzerland (with his mistress) and emerged three weeks later with the discovery of wave mechanics firmly in hand. Albert Einstein was 37 when he formulated the general theory of relativity. Max Planck, the "father" of the quantum revolution that ushered in "boy physics," was 42 when he formulated the theory of quantum energy that started the revolution in physics. Wilhelm Conrad Röentgen, the first recipient of the Nobel Prize in physics, was 50 when he discovered X-rays. Carl Ferdinand Braun was 47 when he developed the cathode ray tube; Dorothy Hodgin was 59 when she solved the structure of insulin. And John Bardeen, the recipient of two Nobel Prizes in physics, was 49 when he and his coresearchers formulated the BSC microscopic theory of superconductivity.

In this chapter we explore reasons that an age–productivity relationship may exist, and in Chapter 4 we look at evidence concerning this relationship. Two questions are of particular interest to us: First, are major breakthroughs more likely to be made by younger scientists than by older scientists? Second, does the productivity of journeymen scientists depend on age? Breaking the question into two parts allows us to distinguish between what might be thought of as two categories of scientists. The eminent—who are exceptionally rare—do the path-breaking work that alters the frontiers of science. By contrast, journeymen scientists do competent, useful research, some of which may be highly creative but is generally not of a nature to alter significantly the course of science. Rather than structure our discussion around these two categories of scientists, however, we examine the reasons that an age–research profile may exist for any scientist, regardless of the scientist's status in the scientific hierarchy. In the conclusion to the chapter we then comment on arguments particularly relevant to the elite and to the journeyman.

WHAT DOES AGE MEASURE?

Before beginning, we should note that age is a measure not only of the amount of time that has elapsed in a life but also the amount of time remaining to live. In a similar way, age measures the amount of time that has elapsed in a career as well as the amount of time remaining in the career or, as phrased in a gaming metaphor, the amount of time left on the clock.[10] For illustrative purposes, consider scientist Altmeyer, who receives a Ph.D. at 28. At age 30, Altmeyer has a career history of two years and a likely future of thirty-five. At age 40 Altmeyer has accumulated a twelve-year career history and has twenty-five more years of career to look forward to. At 46 we find Altmeyer at the halfway point at which the past just equals the likely future. At 55 Altmeyer's past greatly exceeds Altmeyer's future.

Is Altmeyer at 30 more or less productive than Altmeyer at 40, at 50, or at 60? Does a longer career history give Altmeyer an edge? A disadvantage? Does a shorter future spur Altmeyer to play harder in the fourth quarter or lead Altmeyer to relax, having decided the effort is not worth the reward? Do Altmeyer's cognitive processes change with age? Here we examine these questions, drawing on work by sociologists, economists, and psychologists. The reader is forewarned that much of our discussion is speculative and that the age–productivity, age–creativity relationship that we suggest is not inevitable for all scientists. Clearly, age—and the social, economic, psychological, and physiological changes that accompany aging—is but one of many factors that influence scientists' careers.

AGE AND THE WILL TO DO SCIENCE

We argued in Chapter 2 that three motivating forces lead scientists to do science. For want of a better expression, we refer to these as *puzzle, ribbon,* and *gold*. Scientists do science because they enjoy the research process itself. They like the puzzle. They also do research because of a desire to be recognized as the first to solve the puzzle. Scientists like recognition. In addition, as we pointed out in Chapter 2, ribbon is often accompanied by gold. Scientists are also motivated to do research by the desire to earn money. Whether or not scientists spend the same amount of time doing research throughout their careers depends in part on what happens to the motivating forces of puzzle, ribbon, and gold as the scientist ages. Here we focus primarily on the ribbon and the gold. Although we comment on the puzzle aspect, as we shall see, there is less reason to expect an age–puzzle relationship than an age–ribbon or age–gold relationship.

Age and the Quest for Ribbon

By the time scientists take their first professional position, they have a fairly good idea of what is required to gain professional recognition. A necessary condition for success is to publish in highly respected journals, to produce what some would refer to as "base hits." What scientists do not know at the time they leave graduate school is just how good they personally will be at the research game. In addition to effort, there are other determinants of success. Two, talent and luck, are particularly important. Just as some musicians have a special gift that sets them head and shoulders above other musicians, some scientists have an innate ability, a "magic gland," that gives them a special edge. Luck also enters the picture, and in a variety of forms: a serendipitous occurrence in the lab, a referee partial to a particular scientist's research, a competitor who has a dry run in the lab.

A plausible assumption is that at the outset of their careers scientists believe they have a high probability of success, a high probability of having their research judged worthy and published in an important journal. For the purposes of our discussion, suppose that Altmeyer believes this probability to be 70 percent.[11] This is close enough to 100 percent to make Altmeyer work hard, but not close enough to make Altmeyer think that just any scientist can do the research. Suppose that Altmeyer selects a question to study, organizes the research, and performs the experiment. Suppose the experiment fails. Altmeyer tries again, designing a new experiment. Altmeyer is successful and submits the results to a prestigious journal, only to find that a competitor, working one thousand miles away, obtained the same results and submitted an article to the same journal two weeks earlier. Things are not looking good for Altmeyer. Altmeyer changes focus, begins a new experiment, and spends six months in the lab, only to find that a basic assumption, necessary for the experiment to be a success, has been proved invalid. Three research projects, no acceptances. Surely it is time for Altmeyer to reevaluate the probability of doing research that can be published in a top journal.

The scenario for Altmeyer could, of course, be different. The first experiment could be successful and be accepted by a prestigious journal. This could lead Altmeyer to do additional work, also published in top journals. By this time, with several significant pieces of research, Altmeyer begins to have a small following. Altmeyer is invited to conferences and learns who is working on what research questions, a great help if Altmeyer is doing research in an area where the competition is strong. Colleagues begin to refer to Altmeyer's work and defer to Altmeyer as an expert. One consequence of this is that when Altmeyer writes another paper, and the referee is somewhat doubtful about certain points, the editor of the journal decides that Altmeyer is correct: After all, Altmeyer is an expert. Another consequence is that when Altmeyer applies for research funds from a federal agency, the review board is quick to approve the project. Altmeyer is, after all, a "proven" product.

Clearly, the longer Altmeyer works, the better idea Altmeyer has of the probability of doing research deemed worthy of a top journal. In the worst-case scenario, Altmeyer begins to wonder whether 70 percent is not a gross overestimation. Not only does the process of failure and rejection lead Altmeyer to reevaluate the odds of success; the process also makes Altmeyer question whether or not winning is all that important. As a result, Altmeyer is motivated to spend less time doing research. Altmeyer finds, as Altmeyer's career unfolds, that research is not all it was cracked up to be in graduate school.

In the best-case scenario, Altmeyer learns that for Altmeyer, 70 percent is a low probability of getting a "base hit." Success makes Altmeyer spend even more time in the lab. Why? All those requests for preprints and invitations to conferences make Altmeyer even more interested in recognition. Altmeyer becomes a connoisseur of recognition. Altmeyer begins to dream of honors heretofore thought unimaginable. Perhaps one day Altmeyer will receive a major award or honor, such as election to the National Academy of Sciences.

Sociologists of science refer to the process that we have been describing as cumulative, or accumulative, advantage.[12] In particular, they argue that because of this process the recognition awarded to research may increase with the amount of successful research that the scientist has already done. From the point of view of journal referees and the scientific community, research that is really a base hit may look more like a home run. Not only are papers more likely to be accepted; funding also is easier to get. Furthermore, some sociologists contend that a reinforcement process is at work, making the satisfaction derived from additional recognition an increasing function of recognition already awarded.

The Matthew effect, articulated by Robert Merton and named after a passage in the Gospel according to Matthew, states that the recognition value awarded to a scientist's accomplishments depends on the scientist's status in the scientific community. In Merton's words, the Matthew effect is "the accruing of greater increments of recognition for particular scientific contributions to scientists of considerable repute and the withholding of such recognition from scientists who have not yet made their mark."[13] As a result, "two publications of equal intrinsic merit will receive differential recognition if the authors are unequal in prestige."[14] According to Merton, this effect is a result of the vast volume of scientific material published each year. Given this overload, scientists choose their reading material on the basis of the author's reputation, often enhancing the scientist's reputation in the process. Successful scientists not only accumulate advantage. They also acquire a taste for success. A reinforcement process is at work. One researcher told us, "I've been doing research ever since I saw my name in print."

The implications for productivity implicit in cumulative advantage and reinforcement theory are clear: For the individual, scientific productivity is correlated over time. Scientists productive in an early period are productive in later periods; those not productive at an early date are less likely to be productive at a later date. Stated differently, early in their careers all scientists, given their training, do research. Some succeed. Some fail. Some succeed for a while, only to fail later. Success breeds success. Consequently, those who enjoy success continue to be productive throughout their lives; those who have less success become discouraged and eventually look to other pursuits for satisfaction.[15]

Generally speaking, studies confirm this pattern. In a study of physicists, Jonathan Cole and Stephen Cole found that later productivity is heavily influenced by recognition of early work.[16] Physicists who received a great deal of recognition in the form of citations continued to be highly productive, whereas those who received few citations became less and less productive. Similarly, in a study of chemists Barbara Reskin found that early publication and citation contributed to productivity in the next decade.[17]

These results, of course, could also be explained by the "sacred spark" hypothesis that "there are substantial, predetermined differences among scientists in their ability and motivation to do creative scientific research."[18] In statistical terms, there is what is called a *heterogeneous population*: Those with the spark always are productive. Those without it, however, never see their careers take off and flourish. Thus, it is not recognition and the way in which recognition is awarded in science that leads to productivity differences but, instead, a differential distribution of talent within the scientific community.

No one would argue that talent is equally distributed among scientists. A common theme in science is the "special gland" argument. One question, however, is whether the differential distribution of talent accounts for the inequality in research output or whether the inequality can be mainly attributed to the social forces of cumulative advantage. Two studies—one done by Paul Allison and John Stewart in the 1970s and another done by Paul Allison, Scott Long, and Tad Krauze in the 1980s using a more sophisticated research technique—give credence to the cumulative advantage hypothesis without totally discrediting the sacred spark hypothesis.[19] Stephen Cole's work concerning mathematicians is also consistent with the cumulative advantage concept.[20] Furthermore, there is just too much anecdotal evidence concerning unequal access in science to believe that it is all "spark" and no "advantage." For example, a physicist who has held academic positions at several institutions of different quality wrote: "I can tell you that there is a world of difference between writing a letter on Harvard stationery and writing on _____ stationery. In the former case, the door is opened immediately and you get a hearing. In the latter case, you have to knock the door down."

Life's Passages

Altmeyer is not only a scientist. Altmeyer is also a human being. To quote Anne Roe: "Scientists are, first and last, human beings and are subject to all the vicissitudes to which other human beings are subject."[21] Clearly, other things besides recognition and puzzle solving give Altmeyer satisfaction and compete for time, things such as family, community, travel, and leisure activities in general. Early in Altmeyer's life, Altmeyer may feel that giving one's all to science is rational and necessary, both because of Altmeyer's commitment to science and because of Altmeyer's dream of being inducted into a science hall of fame. But as Altmeyer ages, Altmeyer may find satisfaction in other dimensions of life. Does Altmeyer the human go through changes with age that lead to a refocusing of priorities, a reallocation of time?

A popular view is that this indeed occurs, that not only children but also adults pass through various stages as they age. In fact, although the concept of developmental stages began with the study of childhood, with Carl Jung and Erik Erickson the theory moved squarely into adulthood. These theories of adult development, we might add, did not fall on deaf ears but were embraced by a public who became devotees of the developmental concept, as the phenomenal success of Gail Sheehy's 1974 book, *Passages*, indicates.

Developmental psychologists see adults as going through distinct stages of development as they confront a sequence of developmental tasks. Personality development in this perspective is viewed "as a process of individual adjustment to ontogenetic changes in functioning (or need states) that occur among all people."[22] Sociologists and social psychologists also describe adults as going through stages of development but regard these stages as historically specific rather than universal and as based on social expectations and cultural meanings rather than biological maturation. In this social perspective, a role (such as parental, occupational, or marital) may be viewed as having a "career," in which a sequence of events fosters a sequence of changes in behavior and self-image. Furthermore, the force of social expectation makes such patterns appear to be age related and hence natural rather than socially structured.[23]

Here we do not wish to debate the nature-versus-nurture view of stages of development. Instead, we wish only to ask what these stages mean for scientists in particular. Of most interest to us is the stage of development that people report experiencing in their late thirties and early forties. This is a stage characterized by reassessment, a stage that Daniel Levinson calls Mid-life Transition.[24] It is a stage that Bernice Neugarten, in a well-known study of middle-aged persons, found characterized by "reflection" and "stock-taking," a stage in which people structure and restructure their experiences in light of what they have already learned.[25]

Central to the idea of adult development is the concept that there are various aspects of the self or role identities. At certain stages of life, individuals focus on particular aspects, ignoring others, whereas at other stages, previously neglected aspects come to the forefront. In early adulthood, individuals are motivated by the desire to succeed in their chosen occupation, and they measure success in concrete terms. Somewhere in their late thirties or early forties, they begin to reevaluate their life, examining parts of the self neglected while establishing a career. It is a period in which

> it becomes important to ask: What have I done with my life? What do I really get from and give to my wife, children, friends, work, community—and self? What is it I truly want for myself and others? What are my real values and how are they reflected in my life? What are my greatest talents and how am I using—or wasting—them? What have I done with my early dreams and what do I want with them now? Can I live in a way that best combines my current desires, values, talents, and aspirations?[26]

The effort to reevaluate life often involves "de-illusionment" as individuals realize that to date much of their lives has been based on illusion.

Much of this reevaluation takes place when people realize that the "dream" of their twenties is unlikely to be fulfilled. Reevaluation, however, is not only for the unsuccessful. The successful also reevaluate, asking whether there is not more to life than the job to which they have dedicated themselves for so many years. To use Bernice Neugarten's term, this is a period of growing "interiority." It is a period with decreased emphasis on assertiveness and mastery of the environment, a period when one begins to enjoy the process of living more than the attainment of specific goals.[27]

Part of the reevaluation stems from the fact that people in this age category begin to recognize their own mortality, a recognition occasioned by physiological changes in themselves and their friends. Although many of these changes are merely cosmetic (baldness, wrinkles, gray hair), the changes are sufficient to make people realize that their youth is fading. This observation is reinforced as the individuals collect experience with illness and death. As one middle-aged respondent told Neugarten, "There is now the realization that death is very real. Those things don't quite penetrate when you're in your twenties and you think that life is all ahead of you. Now you know that death will come to you, too."[28]

The recognition that they are mortal leads people to restructure the way they perceive time. Until mid-life, it is common for people to measure time in terms of years since birth. But in the middle years of life, usually in their fifties, people begin to think about time in terms of years left to live.[29] It is also interesting that it is approximately at this stage of life that the rate at which people save money increases dramatically.

How does this affect scientists? What does the developmental view portend for them? One possible scenario is that the concept of giving one's all for science seems rational to young scientists. The young love science, and they dream of recognition. The young have been socialized in graduate school into "believing" in science. Young scientists think that they will live, if not forever, then almost forever. But somewhere in their late thirties or early forties, scientists begin to realize that they, too, are mortal. They begin to question whether recognition is worth the extreme effort it requires. They may begin to see their legacy to future generations more in terms of administering programs and teaching classes than in doing research that, if they are lucky, only a few persons will ever read.[30] When offered a position at a "better" institution, they may decide to stay put, choosing life-style over professional recognition. For example, a biologist studied by Levinson and his colleagues declined a chair at Yale at the age of 41 because of strong ties to family and friends and a love of place. His wife is reported to have remarked, "The kudos almost got him, but now we are both glad we stayed."[31]

This reevaluation, of course, does not mean that all scientists in middle age reduce the amount of time they spend on science or the importance of research in their lives. Some may enjoy the puzzle-solving aspects of science so much that they continue at a fast and furious pace, despite competing interests and the knowledge that the clock is ticking. Others may persist because they find the opiate of success to be so strong. Sociologists of science Harriet Zuckerman and Robert Merton suggest that the shift from the research role to other roles is more characteristic of journeymen scientists than of more accomplished scientists. Indeed, Zuckerman and Merton argue that productive scientists "tend to persist in their research roles, forcing death rather than retirement to spell the end of their research careers."[32] The statistician Gauss left a large number of unpublished manuscripts at the time of his death at age 78. (His involvement with his work was so intense that he found it difficult to stop, even for his own wife's death, and reportedly muttered, upon being notified that she was dying, "Tell her to wait a moment till I'm through.")[33] Zuckerman and Merton suggest that this tenacity occurs not only because the successful have accumulated advantage and acquired the taste for recognition, but also because they see this

Why Age May Matter

as a validation process. The extremely successful scientist—the Nobel laureate—may continue to work after receiving the prize in order to "validate the judgment that the eminent scientist has unusual capacities and to testify that these capacities have continuing potential."[34]

Age and the Quest for Gold

Research is rewarded with recognition and also with increased compensation. In Chapter 2 we pointed out that although scientists may not suffer from gold fever, they are not immune to the rewards of money. Does the monetary incentive to do research decline with age? Is it in Altmeyer's financial interest to be more productive at 30 than at 40 or at 50?

Suppose that Altmeyer at 30 makes a significant contribution to science, not earth shattering, but good. Colleagues begin to cite the work. Altmeyer is invited to participate in conferences. The chairman is impressed, the dean is impressed. At raise time Altmeyer gets $1000 more than if there had not been a base hit. But the $1000 is not a bonus; it is a raise. Accordingly, Altmeyer can look forward to getting more in each of the additional 35 years Altmeyer plans to work. Furthermore, because the raise is built into Altmeyer's base salary, future raises, at least cost-of-living increases, will be based on a $1000-higher base salary. What is the monetary value of this research to Altmeyer? Using the concept of present value and assuming an interest rate of 7 percent (compounded annually) and an inflation rate of 4 percent, the answer is about $22,500. Stated differently, the raise ($1000, adjusted for changes in the cost of living, to be received each year for the next 35 years) is equivalent to giving Altmeyer a lump sum of $22,500 upon publication of the article. What if Altmeyer had been 40 instead of 30? Would it have been worth as much? Almost, but not quite. In this instance, Altmeyer would have only 25 years in which to collect the benefits of the raise, not 35, and the present value would be approximately $18,200. Altmeyer might be surprised that the difference is only about $4000, but the reason, as Altmeyer's colleagues in finance patiently explain, is that the 10 years are so far in the future that it takes only about $4300 invested at a 7 percent rate of interest at age 30 to pay out an inflation-adjusted $1000 over the "lost" 10 years. How about Altmeyer at 50? What would be the monetary value of publishing the article then? Now the article would be worth about $12,400, approximately $10,000 less than when Altmeyer was 30, $6000 less than when Altmeyer was 40. By 55 the value of the article would have begun to fall even more rapidly, being worth only $8800, and by 60 the article would have a present value of only $4700.[35]

What must Altmeyer give up in order to produce the research? Time. Early in Altmeyer's life, time devoted to research may not be very costly. The institution expects Altmeyer to do research. The institution "gives" Altmeyer time for research. Besides running a lab and writing grant applications, Altmeyer has few other professional responsibilities. Later in Altmeyer's career, time devoted to research may be more valuable. In economic terms, Altmeyer's time has a higher opportunity cost. If Altmeyer has enjoyed reasonable success, Altmeyer has the opportunity to consult, to give lectures. Altmeyer may also go into administration, where there is less time for research. If Altmeyer has achieved eminence, Altmeyer may have the opportunity to run a major organization such as a university or research center. Research time, of course, is taken not only from professional duties and opportunities; it also comes from time spent with family and friends. The developmental approach suggests that early in Altmeyer's life this time may not appear to Altmeyer to be especially valuable. Later, during Altmeyer's period of reevaluation, the personal cost of research may appear larger, more striking.

To economists, activities that have current costs and generate benefits over a significant period of time must be considered in an investment framework. Rational individuals will purchase less of the investment if either the cost of the investment goes up or the present value of the benefits goes down. Economists do not see this type of perspective as applying only to the decision to build a factory, plant trees, or put wine in the cellar. Economists also see the investment framework as applying to persons making decisions concerning whether to forgo current opportunities in order to earn more in the future. Typically, the type of decision that economists model is the decision to go to school. Economic models that analyze schooling (or on-the-job training) decisions are referred to as *human capital models*, in contrast with what one could think of as *physical capital models*.[36]

Research activity can be modeled in the same type of framework, and several economists have done precisely this.[37] For the individual scientist there are costs associated with research: Opportunities for either earnings or pleasure must be forgone in order to produce the research. Research also generates financial benefits over a period of time. In the human capital framework, scientists will do less research if either the costs of research increase or the present value of the benefits of research decreases. The latter clearly depends on the length of time remaining in the individual's career. As a result, economic models predict that older scientists produce less research than younger scientists do, since each year of age brings with it a shorter and shorter time period over which to collect the rewards of their efforts. This decline will be accelerated if the costs of research increase with age, a distinct possibility, because older scientists have had longer to establish a reputation and therefore may have opportunities to consult and lecture that are simply not available to younger scientists. Older scientists may also value their nonwork time more highly than younger scientists do, a plausible idea given the work of the developmentalists that we just mentioned.

Does Altmeyer make these calculations? Does Altmeyer run the numbers before starting a research project? Probably not, especially when Altmeyer is young and thinks that life will go on forever. And from an economic perspective, this may be rational, for as we have seen, during the early years of Altmeyer's career the monetary value of the research declines little with age. If Altmeyer is attracted by the financial rewards of research at 28, Altmeyer should be almost equally attracted at 35 and at 40. Somewhere in mid-life, however, Altmeyer begins to realize that life is not forever. Altmeyer begins to count not in terms of years out, but years left. Altmeyer begins to think more as the economists predict that Altmeyer will think. Altmeyer may not discount, but Altmeyer may begin to wonder whether the effort is worth the reward. The time to collect the benefits is running out, and the costs of doing research, from Altmeyer's perspective, may be rising. Interestingly enough, this period of reassessment often comes at about the same time that the payoff of doing research begins its precipitous decline.[38]

The fact that Altmeyer, at least at first, may not make these kinds of calculations does not mean that early in the career Altmeyer is not cognizant of the economic benefits of research. Indeed, this is a time when Altmeyer is perhaps most sensitive to the benefits of research, not because of the direct effect of research on earnings, but because of the effect that research has on tenure and promotion.[39] If a young Altmeyer is to continue to have a job in science and be promoted, Altmeyer must do research. Otherwise, there is no hope of getting tenure. After promotion to associate professor the cycle begins again, this time for full professor rather than associate, although at least this time job security is not at stake.

Thus, two economic factors lead to the prediction that research effort will eventually fall with age. Early in their careers, scientists are motivated to do research in order to get tenure and be promoted. This is also a period in their lives when the direct financial rewards to doing research are large and fairly stable and the costs of doing research are relatively low. Later in their careers, scientists are less financially motivated to do research. They have already achieved tenure and been promoted. At the same time, the present-value formula begins to show that with each additional year the rewards to research decline. And it is just about at this time that scientists begin to think in terms of the number of years left in their career.

Age and the Lure of the Puzzle

Scientists do science not only for ribbon and gold. Many also do science because of their love of the puzzle. The neuropsychologist Richard Gregory, for example, says that for him the science puzzle provides more pleasure than other types of games: "I'd rather try to think out a puzzle in my experiments . . . than play a game of chess."[40] Such sentiments are not unusual. It is not just that the puzzle is fun; the solution to the puzzle is itself a reward. Indeed, for some scientists, the solution is the major part of the reward. Richard Feynman reportedly stated in his acceptance speech for the Nobel Prize: "I had already received my prize in the pleasure I got in discovering what I did."[41] For Feynman, the Nobel Prize was anticlimactic.

What happens to the lure of the puzzle with age? Does the satisfaction Altmeyer derives from working the puzzle grow larger; does an older Altmeyer find it easier to play at science? Or does Altmeyer's satisfaction from doing the puzzle grow smaller with age; does Altmeyer's instinct to play diminish? To the best of our knowledge, these are questions rarely discussed, either by scientists or by persons who study scientists. Yet the relationship between age and the puzzle drive is important, as the puzzle is one of three elements, the others being ribbon and gold, that provide scientists with the will to do science. From our study and discussion with scientists, we believe that the puzzle drive does not remain constant with age. Whether it decreases or increases, however, is open to speculation.

According to the developmentalist perspective we described, interest in intrinsic rewards, such as the puzzle, rises with age. This occurs not just because persons become more interiorly motivated. It happens also because unlike ribbon and gold, the puzzle is a reward over which the individual player has control. One wins, in essence, just by playing the game. Thus with age, scientists may find the solace provided by the puzzle to be increasingly important, particularly if they have accumulated little advantage over their careers. Another reason to believe that scientists may become more interested in the puzzle aspects with age relates to the fact that for some scientists science approaches being a religion. Such scientists perceive there to be truth in nature and use their tools as scientists to find this truth. For such scientists, the religious aspects of science may become increasingly important with age, as they become more aware of their own mortality. Einstein in later life, for example, devoted a great deal of energy (unsuccessfully) to "the holy grail: the unification of gravity with other forces in nature."[42]

It is also possible that the puzzle loses much of its charm with age. Several factors lead one to think this may be the case. Two are particularly noteworthy. First, it is possible that because of the way science is structured in the United States, the fun of the puzzle lessens with age and what was once play becomes a chore. The dominant role played by grants in U.S. science makes this argument particularly plausible. In today's world, by the time scientists reach middle age—and in many instances long before that—the pressure to obtain grants has become formidable. Scientists are expected to bring in dollars for themselves, their labs and also for their graduate students. Increases in pay, the gold, may depend as much on the amount of funding brought in as on the amount of research performed. Funding pressures may turn play into work, especially if only safe projects, low in imagination but high in predictable results, can get funded. Scientists become bored with such problems, and so they are no longer play. And even if the funding is for "play," for things the scientist loves to do, the scientist may become overwhelmed by the enormous amount of time that grantsmanship demands and the constancy of its demands. Many hours are needed to complete grant applications, and success brings even more hours of work writing progress reports and final reports. The grantsmanship calendar has a life of its own. A 39-year-old biologist told us that she skipped Christmas Eve festivities because she was hard at work writing a grant proposal, and a clinical physician left town the day after Christmas to review grant proposals. Such sacrifices make scientists wonder whether science is still fun.

One other reason leads one to speculate that the attractiveness of the game ay diminish with age. Particularly for scientists who have not experienced a deal of success, the period of interiority that Levinson described may be nied by a feeling of alienation from their field, a feeling that the sociologist Hagstrom relates to anomie.[43] Scientists may begin to wonder whether it is really worth spending time on the puzzle if virtually no one besides themselves appreciates the solution. Puzzles are fun, but it is even more fun to share the solution with others. With age, many scientists — even scientists who have published with some regularity — may become increasingly aware of the fact that the majority of scientific papers are cited by at most one or two people and — one can only infer — go largely unread. Anomie may set in.

AGE AND COGNITIVE RESOURCES

Scientists give more than time and effort to research; they also provide cognitive resources. Thus, in addition to questioning whether the will to do science changes with age, we must also question whether the cognitive resources of the scientist change with age. In terms of the framework used in Chapter 2, these cognitive resources have two dimensions: the scientist's knowledge base and the scientist's innate ability to do science. Although the two are clearly related in the sense that ability affects the learning process and hence the knowledge base, to facilitate our discussion we shall examine each separately.

Age and the Knowledge Base

To a large extent, Altmeyer's research efforts are determined by what Altmeyer knows. We can think of this dimension of Altmeyer as Altmeyer's knowledge base, a large part of which was acquired in graduate school. Here we discuss what happens to this knowledge base as a consequence of the aging process. The related issue of whether a scientist's knowledge becomes obsolete is considered in Chapter 6, because obsolescence is not related to the aging process per se but instead to what happens in the external environment as the scientist ages.

Economists argue that early in Altmeyer's career Altmeyer's knowledge base expands, as Altmeyer has the incentive to learn more. This kind of reasoning is based on the human capital framework mentioned earlier in this chapter and assumes that the individual views learning as a conscious investment that has both costs and benefits. The costs are primarily composed of forgone opportunities. Learning takes time: time to read preprints and journals, go to conferences, talk with colleagues. Time for learning also means less time for other things. In economic terms, time costs money. Thus, in deciding whether to learn more, Altmeyer is assumed — according to this economic model — to weigh the costs of learning against the benefits of acquiring new knowledge. In the economic perspective, these benefits are thought to diminish with age, for as scientists age they have fewer and fewer years to use their new knowledge to produce income-generating activity.

Economists extend the analogy between human capital and physical capital even further. They argue that persons make conscious decisions about how much to learn. But they also argue that human capital, like physical capital, depreciates. Here, however, the analogy between physical capital and human capital comes up short, because knowledge does not wear out through overuse but instead may fade as a result of underuse. For example, mathematical skills acquired in graduate school, but left unused, diminish with time. Thus as Altmeyer ages, Altmeyer both acquires new knowledge and "loses" some knowledge and skills learned at an earlier time. As long as Altmeyer acquires more knowledge than Altmeyer loses, Altmeyer's knowledge base increases. But according to the human capital theory, because the incentives to learn decline with age, a time may come in Altmeyer's career when the amount of new knowledge acquired is less than the amount of knowledge lost as a result of disuse. If such is the case, the 55-year-old Altmeyer may know less science than the 35-year-old Altmeyer did.

Although this modeling of the learning process is quite superficial (Does Altmeyer really calculate whether Altmeyer should read one more preprint, attend one more conference?), it captures several aspects of the process that are worth noting. First, much of the knowledge that scientists acquire after graduate school comes through experiencing science. Thus, as long as they spend time doing research, time talking with colleagues, scientists continue to learn. Second, because scientists do not use some acquired skills on a regular basis, after a while they become rusty. Furthermore, if the incentives to do science decline with age, as we have argued may be the case, it is quite possible that scientists also acquire less new knowledge as they age. And because knowledge and research skills not used on a regular basis are likely to fade, it is not unreasonable to think that at some point older scientists may know less than younger scientists do. Unless the "fade-out-rate" is very high, however, this is not likely to happen until the scientist's career is fairly advanced. Age can, on the other hand, give scientists a leg up in knowledge acquisition: Older scientists may possess a better framework for storing knowledge and be more efficient and/or selective in choosing what to learn.

Age and Mental Processes

Altmeyer's cognitive resources also depend on the functioning of Altmeyer's brain. Is this functioning related to age? Does Altmeyer's ability to do research change as Altmeyer ages?

The brain, in Barash's metaphorical language, is "three pounds of very soft white cheese, wavy like a cauliflower; crisscrossed by electrical circuits that beggar the wildest imaginings of a Bell Telephone engineer; a juicy gland whose chemical drippings we are just beginning to identify, never mind to understand; home to all our thoughts, memories, passions and fears."[44] That this mass of cheese changes with age is now fairly well accepted. One example is the loss of nerve cells, estimated at approximately 100,000 in twenty-four hours.[45] Such a loss prompted Francis Braceland, a past president of the American Psychiatric Association, to say with regard to age, "When it's quiet I can almost hear brain cells drop out,"[46] clearly an exaggeration, as the yearly loss amounts to only .04 percent of the total number of nerve cells in the brain.

It is also well documented that the electrical activity of neurons in the brain changes with age. This activity is studied by attaching electrodes to an individual's scalp. The pattern of electrical activity recorded is called an *electroencephalogram* (EEG).[47] There are at least four distinct EEG rhythms. The dominant brain-wave rhythm, the alpha rhythm, has a range of 8 to 13 cycles per second (cps). In regard to behavior, alpha waves are associated with a relaxed but alert state. A rate above 13 cps is referred to as beta activity and is associated with alert thinking. Theta activity, in a range of 4 to 7 cps, is associated with drowsiness and daydreaming, and delta activity, in a range of 1 to 3 cps, occurs during sleep. For young adults, the dominant brain-wave rhythm is between 10.2 and 10.5 cps. "One of the best-documented findings in the psychophysiological literature is that this dominant brain-wave rhythm slows with age. By the time an individual reaches the age of 60–65, his or her dominant brain-wave rhythm is probably around 9 cps."[48] Furthermore, this result is found even in very healthy older people.

Another conclusive finding is that a slowing of behavior accompanies aging. For over twenty-five years the gerontologist James Birren has maintained that a slowing of behavior is the most pervasive age-related change.[49] This finding is so robust that psychologists are at a loss for new adjectives to describe its reliability. The slowdown occurs across a wide spectrum of behavior, from zipping a garment to dialing a telephone, unwrapping a Band-Aid, cutting with a knife, and even putting on a shirt. The higher accident rates of older adults are usually linked to a slowed response-time, and the underrepresentation of older adults in externally paced industrial tasks has also been linked to the slowness of behavior that comes with aging. The evidence also suggests that the slowdown of behavior, at least of some behavior, begins in the mid-thirties, long before solid middle age sets in.[50]

Memory is another brain-related activity for which there is almost universal agreement that an age relationship exists. "The feeling that one's ability to remember and to retrieve information is not as good as it used to be is a universal complaint among middle-aged and elderly persons."[51] The finding of a relationship between age and the ability to recall information is so robust (and so believable for anyone past 40) that researchers in recent years have focused on explaining why the relationship exists rather than questioning whether it exists.

Memory studies generally distinguish among several stages of information processing. *Sensory memory* is a term used to refer to the stage in which new information is initially registered. *Primary memory* is perceived as a separate stage, having limited capacity, in which information still "in mind" is kept as it is being used. *Secondary memory* refers to the stage in which information not currently in use is stored. It is this memory concept that the general public thinks of when speaking of memory, and it is this memory component that shows a definite age decrement. In the words of two prominent memory researchers, "Aging . . . exerts a profound effect in the acquisition and retrieval of new information in secondary memory."[52] Although there is no consensus as to why the decrement exists, Leonard Poon believes that there is reasonable evidence to support the idea that as people age, they become less adept at organizing material and hence possess storage systems inferior to those of younger persons. This hypothesis is consistent with the finding that highly verbal people suffer less memory loss with age than do less verbal persons.[53] It is also consistent with the idea that as Altmeyer ages, not only are there fewer incentives to learn, but also the capability to learn may decline.

What does this mean about ability and age? Does Altmeyer become less able with age? Does a decline in alpha waves, a slowing in behavior, and a loss of memory add up to an age decrement in IQ? The conventional wisdom among specialists concerning this question today is no.[54] Early studies, however, generally concluded that an age decrement existed and that it began at a fairly early age. Before looking at the conventional wisdom of today, we shall briefly review these studies. One reason we pause to do so is that the methodological issues encountered in examining the age–IQ relationship are similar to those encountered in examining the age–productivity relationship, which we discuss in Chapter 4.

Before the 1960s, studies of intellectual development in adulthood used cross-sectional data. That is, in these studies, adults of various ages were given IQ tests at the same point in time. A consistent finding in these studies is that ability peaks in young adulthood and then diminishes with age, although the degree of age decrement varies with the ability being tested.[55] Researchers, however, began to wonder whether the effect was due to age per se or to other variables related to age when they noted that the age of peak performance reported on IQ tests had increased over time. (Studies done in later historical periods reported later peak ages than did the studies done at earlier times.) This led psychologists to question whether the results might be generation dependent. That is, in a test given in 1950, older persons may have scored lower than younger persons did, not because they were older, but because they came from an earlier generation that had access to less education or less up-to-date information. Beginning in the 1960s Warner Schaie and others began to approach the IQ–age–decrement question using a longitudinal methodology rather than a cross-sectional methodology.[56] This meant testing the same individuals over time as they aged. The major benefit of this type of study is that it permits researchers to separate generational effects (which are also called cohort effects) from aging effects.

The results of these longitudinal studies differ significantly from the results of the cross-sectional studies. Indeed, these studies strongly suggest that at least until sometime in the late fifties or early sixties, age has virtually no effect on IQ, but generation does. That is, Altmeyer does not become less smart as Altmeyer gets older; rather, successive generations of Altmeyers get smarter at least in terms of being more up-to-date. In Schaie's words, "There is very little age decrement in intelligence in functions that do not require speeded response or are not affected by the slowing of reaction time within the individual. . . . [But] marked differences in performance level between successive generations" exist.[57]

Finally, in discussing mental ability we should point out the distinction between physiological changes due to aging per se and physiological changes due to age-related illnesses such as diabetes and cardiovascular disease. We have assumed that Altmeyer is healthy. With age, however, Altmeyer is more and more likely to suffer from certain age-related illnesses, many of which can and do affect a person's mental processes, independent of the effect of age per se. Thus, Altmeyer's mental processes may decline with age not as a direct result of aging but because of the onset of diseases known to be age related. Such illnesses also consume a substantial amount of time and energy. Not only will the sick Altmeyer possibly have a mental decrement, the ill Altmeyer will have less time to devote to science.[58]

AGE AND CREATIVITY

Good scientific work is generally thought of as creative, in terms of both the finished product and the cognitive processes used to arrive at it. Although other distinctions can also be drawn,[59] the process–product distinction is a useful one to make when discussing the relationship between age and creativity.

Creativity as a Process

Creativity is often associated with mental characteristics such as the ability to conceptualize, to solve problems, and to engage in unstructured and divergent forms of thinking. Creativity is also generally associated with intelligence, although high IQ is clearly not sufficient for creativity.[60] In addition to ability, empirical work identifies a set of personality characteristics associated with creative achievement. These include "high valuation of aesthetic qualities in experience, broad interests, attraction to complexity, high energy, independence of judgment, autonomy, intuition, self-confidence, ability to resolve antinomies or to accommodate apparently opposite or conflicting traits in one's self-concept, and finally, a firm sense of self as 'creative'."[61]

Factors affecting creativity can be field dependent. That is, what leads to creativity in literature, poetry, or music may not lead to creativity in science. Consider, for example, the relation between psychopathology and creativity. Whereas certain types of mental illness may foster creativity in fields such as literature and art, there is no evidence that this is true in science. Indeed, the evidence reveals that creative scientists tend to be "more emotionally stable, venturesome, and self-assured than the average individual, whereas creative artists and writers tend to be less stable, less venturesome, and more guilt prone."[62]

Why may creativity be age related? One reason is that some of the cognitive factors associated with creativity are age related. IQ eventually declines with age, although only by a small amount and only late in life. Some research describes older persons as being more cautious and more rigid in their thinking than younger persons are, as well as having greater difficulty conceptualizing and solving problems in general. The current consensus, however, is that these attributes (some of which may not even hold for older persons) are not sufficient to make creativity age related.

On the other hand, there is some support for the idea that age is related to the creative process not because of a decline in the mental processes associated with creativity but because younger persons possess a fresh point of view. They are less encumbered. They do not know about past failures, and they bring a new outlook to solving a problem. Philo Taylor Farnsworth, the youthful inventor of television, is reported to have had success because he "didn't know it wasn't possible." Watson suggests that Linus Pauling failed to discover the structure of DNA before Crick and himself because he resisted (at the age of 50) the idea of a helical structure.[63] This purely anecdotal evidence suggests the existence of a mind-set in mature scientists that keeps them from discovering new truths.

The fresh point of view of young scientists relates to processes used to solve a problem as well as to the choice of the problem to be studied. Indeed, there is much to indicate that truly creative minds see questions that others have not thought to ask. "Many creative individuals have pointed out that in their work the formulation of a problem is more important than its solution and that real advances in science and in art tend to come when new questions are asked or old problems are viewed from a new angle."[64] Such certainly was the case for Newton, Freud, and Darwin.

In the nineteenth century, the psychologist G. M. Beard[65] elaborated a theory of why creativity might be age related. According to his theory, two elements are key to creativity: enthusiasm and experience. The former provides the motivational impetus for ideas, and the latter allows for the effective articulation of ideas. In Beard's view, the young have lots of enthusiasm but little experience, whereas the old have little enthusiasm but lots of experience. Enthusiasm without experience produces original but unfocused work, and experience without enthusiasm produces uninspired work. Thus, Beard predicted that creativity is at its zenith when the two factors cross. One appeal of this approach is that it explains why the age at which creativity peaks may be field dependent.

Beard's belief that experience may be more important in some fields than in others is loosely related to the concept of codification, popular in sociology. The sociologists Zuckerman and Merton, for example, define codification to be "the consolidation of empirical knowledge into succinct and interdependent theoretical formulations."[66] In highly codified fields there is a strong consensus as to what constitutes the important questions and the proper approaches to solving these questions. In less codified fields there is substantially less consensus. Thus, in highly codified fields, experience may matter less because one does not need to be familiar with as many points of view. But in less codified fields, experience may be extremely important.

In this century, the psychologist Dean Keith Simonton has extended Beard's model. Like Beard, Simonton sees creativity as a two-tiered process. But for Simonton the first step is ideation, the transformation of ideas from the potential to the actual, and the second step is the elaboration of these ideas. Simonton posits that ideation is a decreasing function of time in the field, because in his view, the longer someone has been in the field, the fewer are the potential ideas left to be ideated.[67] Elaboration, on the other hand, depends on the number of ideas already ideated waiting to be worked out. Early in the scientist's career, ideas are transformed from the potential to the actual. In the process, the scientist builds up a stock of ideas to elaborate later on. Eventually, this stock grows smaller, and consequently the number of ideas articulated in a given period also begins to wane. Creativity has peaked.

The Simonton model can also be used to explain why creativity may vary across fields.[68] In some fields, ideas can be actuated quite quickly. Einstein knew the precise day on which, "there came to me the most fortunate thought of my life."[69] In other fields, it takes years to articulate an idea. In these highly uncodified fields the age–creativity profile is fairly flat. The contrast between "zombie biology" and physics illustrates this point: The molecular biologist Sydney Brenner once thought of founding a journal called the *Journal of Zombie Biology*, not for the biology of zombies, but for zombie biologists. His reason: "Because that's all you have to do. You just have to wind yourself up in the morning, and go to the lab and just do things . . . many of the answers come from just doing things. Biology isn't a subject in which you can have great thoughts in the bath." The nuclear physicist Leo Szilard (who left physics to work in biology) once told Brenner that he could never have a comfortable bath after he left physics. "When he was a physicist he could lie in the bath and think for hours, but in biology he was always having to get up to look up another fact."[70]

There are also noncognitive reasons why creativity may be age related. Many of these relate to the sociological concept that persons coming from the margin—"outsiders" if you will—make greater contributions than do those firmly entrenched in the system.[71] The margin facilitates a fresh point of view. Moreover, people from the margin have less invested in a particular idea or school of thought, and so they have neither status nor reputation to lose if their research fails. Thus the costs of looking at something from a different perspective are low for "marginal" persons, but the benefits may be extremely high. Because young persons are new to the field and have not accumulated a reputation, they are, in this sociological perspective, marginal and hence more likely to be creative.

One other reason that the process of creativity may be age dependent relates to scientists' belief system. Many scientists firmly believe that science is a young person's game. Therefore, as they age, scientists may not only be written off by their colleagues as "washed up," but they may also write themselves off as not being capable of making a significant contribution. Altmeyer is less creative as Altmeyer ages because Altmeyer believes Altmeyer "lost it" at age 30.

Creativity as a Product

For many scientists creativity is a goal, not just a process. Scientists do not want to do just any research; they dream of doing research that is judged creative, for the accolades of science are tied to the creativity of the accomplishment. Does the conscious goal to be creative lead scientists to be less so as they age? Possibly, because creativity takes time and thus requires effort. As the psychologist David Harrington stated:

> My review of creative episodes has led me to conclude that creative work often requires long periods of sustained, nonalgorithmic interactions with the cognitive and physical materials of the project. . . . Although less glamorous and exciting than the sudden "insights" and "breakthroughs" that fascinate new students of creativity, these periods of nonalgorithmic work appear to be necessary antecedents of "insight" and "breakthrough" in most creative episodes.[72]

Once time is factored into the creativity equation, the reasons given earlier in the chapter suggest that the incentives to do creative work decline, or eventually decline, with age. For example, the cumulative advantage scenario posits that with age, scientists may learn that they are less likely to do creative work than they had originally thought, and consequently they become discouraged from spending time on creative endeavors. By mid-life, they may look to other pursuits for pleasure. Or as scientists age they may, in the human capital framework, "run the numbers" and decide that the reward to hard work is not worth the effort. Furthermore, with age scientists often find that they have less and less time to be creative because they are required to take on other, non-research roles. One such demand is the pressure placed on senior scientists to spend more and more time in "the money chase for grants," as the Nobel Prize winner Leon Lederman refers to it.[73] Younger scientists are less encumbered in a cognitive sense and also less encumbered by the amount of time they are called upon to give to other dimensions of science. One of these, as we have already mentioned, is the need to write grant proposals to support the research team.

Finally, before concluding our discussion of creativity, we should comment on whether the factors motivating creativity relate to the amount of creativity that is forthcoming from an individual. The psychologist Teresa Amabile, for example, argues that intrinsic motivation (in our terms, interest in the puzzle) spurs creativity, whereas extrinsic motivation (interest in ribbon and gold) inhibits creativity. Indeed, she claims that her research results are "sufficiently compelling that we now refer to the intrinsic motivation principle of creativity: intrinsic motivation is conducive to creativity, but extrinsic motivation is detrimental."[74] She goes on, however, to admit that extrinsic motivators need not undermine creativity and that for some people, creativity may actually be enhanced in the face of extrinsic motivators.

In some sense, the intrinsic–extrinsic creativity relationship relates to our concept of encumbrance and marginality. Creative work may be judged creative precisely because it represents a different point of view; it comes from the margin. Viewed in this way, it would make sense that interest in extrinsic rewards is detrimental to creativity if the interest acts to inhibit a different point of view. On the other hand, interest in extrinsic rewards may spur creativity if the interest leads the individual to go outside the system for a solution. If this view is correct, we would expect the relationship between extrinsic interest and creativity to be field dependent. In fields in which financial and perhaps reputational rewards can be achieved by following a somewhat conservative and less innovative path, interest in extrinsic reward may stifle creativity. In other fields, however, in which reputation and financial rewards are bestowed more frequently on an approach coming from the outside, an interest in extrinsic rewards may enhance creativity. We believe that much of science falls squarely in the latter category.

A Head Start, a Nose for Success?

Our discussion of cognitive processes would not be complete without pointing out that youth and significant work may go hand in hand in science because of what could be called processes of selection. For example, scientists who are extremely creative—who have what some refer to as the magic gland for doing science—often begin to make discoveries earlier in their lives than do other scientists. That is, they have a head start precisely because of this ability. Pascal wrote a paper on conic sections at the age of 16 or 17; Galileo discovered the isochronism of the pendulum at 17; and by the age of 22 both Darwin and Einstein were publishing.[75] American Nobel laureates in science, according to Zuckerman's calculations, publish on average more than a dozen papers before their thirtieth year (approximately four times as many as scientists in general contribute in an entire lifetime.)[76] Thus, a relationship between age and creativity may exist because great contributions are made by scientists with the special gland, and such scientists, precisely because they do have a special talent, start doing science at an earlier age than do those who are not as talented. From this it does not necessarily follow that science is a young person's game; exceptionally talented scientists may continue to be productive as they age and are joined by other, less able scientists in their age group.

A related issue is that the best minds in science not only start early but also are attracted to areas of science in which major breakthroughs are occurring (or are likely to occur), for it is in these fields that research will have the highest rewards. That is, they have a nose for success. For example, Crick and Watson chose to do research on DNA because they saw this as being an area on the cutting edge. According to the sociologist Judith Blau, breakthrough fields "attract the most capable young scientists, who in turn, by virtue of superior abilities and training, solve the problems in short order."[77] According to this view, it is not necessarily that young talent is required to solve special problems, it is just that young talent is especially attracted to "hot" fields. After the field has been mined and the scientists are older, they may be content to do backwater or what some call ditchdigging research, rather than to retool and challenge a new frontier. The latter course, after all, is not only risky but also involves substantial costs. Clearly, because the definition of the frontier depends on events in science, Blau's idea is related, at least at the first-cousin level, to the idea of obsolescence and change in science, which we shall discuss in Chapter 6.

CONCLUSION

In this chapter we have explored why scientists' age may affect their willingness and ability to do scientific research. In the introduction, we divided scientists into two groups. What does our discussion have to say about the elite and the journeymen? We would argue that for both groups there is reason to believe that research activity eventually declines with age, particularly somewhere around mid-career when developmental and economic forces combine to make research less appealing, thus lowering the will to do science. For journeymen scientists, this decline may be reinforced by the processes of cumulative advantage. The same processes may soften the decline for elite scientists, and it may be further softened by the desire of the eminent to persist with research in order to prove themselves worthy of the recognition that they have already received. In spite of these mitigating factors, two forces may conspire to make the eminent less likely to be as productive in later life. First, the elite possess some sort of magic gland that, for a variety of reasons, appears to work best from youth to early middle age, particularly if their work is abstract, wherein facts are less important than theory. Second, because of their eminence, elite scientists are often offered opportunities that the noneminent never have. Usually these come in middle age and involve administering something big, such as a university, a laboratory, or a company. Thus, at approximately the same age that the expected benefits to future research become noticeably lower, the opportunity costs of continuing in research may become substantially higher. The eminent may find the career change irresistible, especially when the position provides status, something for which the scientist has acquired a taste.

In the next chapter, we look at evidence concerning these two groups of scientists. In addition to reviewing work done by others, we provide estimates of the age–productivity relationship derived from a unique database that we compiled in the 1980s.

NOTES

1. Brodetsky 1942, p. 699.
2. See Giere 1988, pp. 252–7; Zuckerman 1977, p. 166.
3. Pickering 1984, p. 78.
4. Simonton 1984, p. 99.
5. Wolpert and Richards 1988, p. 20.
6. Lehman 1956, pp. 336–7.
7. Zuckerman 1977, p. 164.
8. Wolfe 1983, p. 371.
9. Lehman 1953.
10. In the social sciences, sociologists generally use age as a proxy for the amount of experience that an individual has accumulated in a career. Economists, though interested in experience, focus on the idea that age is a measure of time left in the career or, more generally, in life. To the extent that behavior changes with age, sociologists, therefore, believe that the change is due to experiences accumulated during the career. Economists, on the other hand, believe it is related to the fact that older persons have a shorter time to realize the benefits of their actions. A similar distinction exists in the life sciences, in which scientists cannot entirely determine whether the effects of aging are

due to the passage of time or the approach to some biological limit. Thus, for example, the physiological aging process may be a result of repeated exposure to radiation or because cells in the body are coded to cease functioning at a certain time.

11. Note that 70 percent is the approximate acceptance rate by journals in the physical and life sciences. Zuckerman and Merton (1971) report acceptance rates of 60 to 80 percent in the physical sciences, and Hargens (1975) cites rejection rates of about 23 percent in physics, 29 percent in the biological sciences, and 30 percent in chemistry. For geology he found a rejection rate of 20 percent. In more recent work, Hargens (1988) reports acceptance rates for major journals for the early 1980s. Although he found variation across scientific fields, among the life and physical sciences the acceptance rates are, in most instances, greater than 65 percent.

12. See, for example, Allison, Long, and Krauze 1982; Allison and Stewart 1974; Cole and Cole 1973; Merton 1968.

13. Merton 1968, p. 58.

14. Allison, Long, and Krauze 1982, p. 615.

15. Price (1976) points out that the Matthew effect described in the Bible is a two-sided sword and that a one-sided sword may better fit the data. That is, success breeds success, but failure does not breed more failure.

16. Cole and Cole 1973.

17. Reskin 1977.

18. Allison and Stewart 1974, p. 596.

19. Allison and Stewart (1974) argue that in order to differentiate between these hypotheses, it is essential to have longitudinal data, since the sacred spark model does not imply a change in the distribution of output over time, whereas the cumulative advantage/reinforcement model may. Lacking longitudinal data, however, they divided a cross-sectional sample of scientists into strata by career age and computed Gini coefficients of productivity inequality by age strata. For both article publications (over a five-year period) and citations (to all published work), they found the evidence strongly supporting the cumulative advantage hypothesis. Allison, Long, and Krauze (1982) tested the cumulative advantage model using longitudinal data for biochemists and chemists and found increasing inequality over time for publication counts but not for citation counts.

20. In a longitudinal study, Stephen Cole (1979) found that over a twenty-five-year period, the percentage who were not productive increased from 38 to 61 percent, and the percentage who were strong publishers remained approximately constant, in the neighborhood of 15 percent, over the same time period.

21. Roe 1963, p. 132.

22. Lawrence and Blackburn 1988, p. 23.

23. See Dannefer 1984.

24. Levinson 1978.

25. Neugarten 1968a, p. 139. Although sociologists and social psychologists question whether the mid-life reevaluation is universal in our society, they recognize that the male "mid-life crisis" is common among professionals and managers for whom work has high salience and who are likely to develop a clear sense of whether they are or are not "on time" in their career progress (see Dannefer 1984).

26. Levinson 1977, p. 107.

27. Levinson 1978, p. 196.

28. Neugarten 1968b, p. 97.

29. Neugarten 1968a, p. 140.

30. It is impossible to know the percentage of scientific papers that are never read. It is possible, however, to ascertain the percent that are never cited over some period of time. A recent study by the Institute of Scientific Information estimates that during the first five years of their life, 22.4 percent of scientific articles are never cited. If one restricts the analysis to U.S. authors, 14.7 percent are never cited (see Pendlebury 1991). Many articles, of course, receive only one or two citations. Among publications cited in *Science Citation Index, 1975-1979*, the median and modal number of citations converge at one (Stern 1990, p. 193).

31. Levinson 1978, p. 267.

32. Zuckerman and Merton 1973, p. 529.

33. Simonton 1984, p. 84.
34. Zuckerman and Merton 1973, p. 532.
35. Many pensions are based on salary. Thus, the rewards to publishing may last into retirement, and the scientist would, therefore, discount over a longer period of time. (Discounting refers to the process by which a stream of income to be received in the future is converted to its present value, that is, converted to the amount of money that, if invested today at the assumed interest rate, would yield the same future stream of income.)
36. For examples of such models, see Becker 1964; Ben-Porath 1967; Schultz 1963; Stephan 1976.
37. See, for example, work by Diamond 1984; Levin and Stephan 1991; McDowell 1982.
38. The speed with which the present value changes over time does, of course, depend on the interest rate as well as the rate of inflation.
39. Tenure and salary are related, but tenure is more immediate.
40. Wolpert and Richards 1988, p. 195.
41. Feynman 1985, p. 281.
42. Pais 1986, p. 30. Einstein did not attempt to unify the four forces. Rather, he struggled to unite gravitation and electromagnetism, the two forces known to him at the onset of his work.
43. Hagstrom 1965.
44. Barash 1983, p. 125.
45. Ibid., p. 126.
46. See Busse 1989, p. 14.
47. For a discussion of brain rhythms, see Woodruff 1983.
48. Ibid., p. 182.
49. See Birren, Woods, and Williams 1980 for a review of this literature.
50. This discussion draws heavily on Salthouse 1985. Some of the first work done analyzing reaction time and age used data collected by Sir Francis Galton in 1877 at a health exposition in London (Woodruff 1983, p. 182).
51. Poon 1985, p. 427.
52. Some researchers identify *tertiary memory* as another stage, a stage at which well-learned and personal information such as childhood memories is stored. There is little evidence that tertiary memory is age related (see Siegler and Poon 1989, p. 171). Tertiary memory is also difficult to study, since researchers seldom have the luxury of observing subjects over a twenty- or thirty-year period.
53. Poon 1985.
54. This answer assumes that older persons have enough time to take a test and that the test is constructed in such a way that it does not depend on reaction time.
55. Schaie 1958.
56. Schaie and Labouvie-Vief 1974.
57. Schaie 1983, p. 145.
58. A study by the National Institute of Mental Health (1971) found significant differences among a variety of factors between healthy older adults and older adults who were classified as having mild asymptomatic or subclinical disease.
59. Simonton (1990) goes further and speaks of the four "P's" of creativity: product, process, person, and persuasion. Here we focus primarily on the first two. Process and person are related, and persuasion, in Simonton's terms, relates to the ability to persuade others of the importance of one's work.
60. High IQ is not sufficient for creativity. Robert Albert's review of studies of the relationship between exceptionally high IQ and creativity leads him to conclude that "more than cognitive giftedness is required for exceptional career achievement" (Albert 1990, p. 17). There is some support for the threshold theory of creativity. "According to this theory, some intelligence is necessary for creative performance but only in the moderate and high levels of intelligence is creativity ostensibly independent. Some have gone as far as to suggest that an IQ of 120 represents the minimum IQ threshold for creativity" (Runco 1990, p. 241).
61. Barron and Harrington 1981, p. 453.
62. Ibid., p. 456. Others have a similar view of pathology. For example, Raymond B. Cattell

(1963, p. 122) stated, "For the present I am inclined to see the general evidence as agreeing with Terman that emotional stability may be low for literary geniuses, *but that the average level of ego strength and emotional stability is distinctly higher for effective scientific researchers than for the general population.*" Simonton (1984, p. 55) also observes that there is a tendency for "artistic creators to display more emotional instability than scientific creators." Simonton also points out that pathological symptoms found in artistic people could be the consequence of creativity rather than its cause.

63. Watson 1968.
64. Einstein and Infeld 1938, p. 92.
65. Beard 1874.
66. Zuckerman and Merton 1973, p. 507.
67. Simonton 1983, 1984. Simonton assumes that each creator begins with a fixed supply of "creative potential."
68. The Simonton model differs significantly from the Beard model in that the Beard model predicts that the quality of work also is age related, reaching its peak when enthusiasm and experience meet in the correct mix. The Simonton model assumes that the ratio of high-quality work to poor-quality work is constant over the career, that what changes is the amount of output.
69. Unpublished manuscript for *Nature*.
70. Wolpert and Richards 1988, p. 107.
71. Gieryn and Hirsh 1983.
72. Harrington 1990, p. 156.
73. National Science Foundation 1990a, p. 5.
74. Amabile 1990, p. 67.
75. See Simonton 1984, p. 84, for examples of precocious scientists.
76. Zuckerman 1977.
77. Blau 1978, p. 204.

4

Age and Scientific Productivity: Must One Be Young to Do Great Things?

Chapter 3 discussed why a scientist's productivity may decline with age. In this chapter we look at the related evidence taken from studies done by other researchers as well as from those done by ourselves.[1] We focus on two aspects of the age–productivity question: the age at which eminent scientists do their best or most noted work and the relationship between age and total productivity for scientists drawn from the journeymen ranks as well as from the ranks of the eminent. We refer to the combined group as *average* scientists and investigate their productivity rather than the productivity of journeymen by themselves because data that include journeymen rarely omit the eminent.

The data used to study productivity issues come from historical records as well as from samples of scientists. When samples are taken, they may be either cross sectional, in which all data are collected at the same time, or longitudinal, in which a group of scientists is followed over time. Productivity is generally measured in terms of either an idea (or the article that embodied the idea) or an invention. As can easily be imagined, research findings often depend on the methods used in the analysis, and thus there is much debate about the correct methodology to use. Rather than bore the reader by examining methodological issues in the abstract, we present the evidence and then discuss the methodological issues as they arise. We begin with the productivity of eminent scientists.

EMINENT SCIENTISTS

Lehman's Work

The largest study of age and productivity of eminent scientists was undertaken in the early 1950s by the psychologist Harvey Lehman.[2] Indeed, the scale of Lehman's work was so massive that it spanned over thirty years of his career and included chemists, physicists, astronomers, entomologists, geneticists, agricultural chemists, bacteriologists, physiologists, pathologists, and anatomists, as well as those making discoveries in medicine and innovations in surgical techniques. In addition, Lehman studied scientists who contributed descriptions of disease and made advances in medicine and public hygiene, and those who introduced drugs and remedial agents. Furthermore, Lehman did not limit his study to scientists but studied as well athletes, poets, composers, master chess players, novelists, and painters.

Lehman approached the question of age and productivity by using standard reference works to identify great scientists and their contributions.[3] He then ascertained the age at which the contributions were made. Restricting himself to scientists deceased at the time of his study, Lehman next computed the average output by age interval for those scientists who lived at least to that age and normalized the average output in terms of the most productive group. A hypothetical example of this methodology is given in Table 4-1 for one hundred eminent scientists making a total of 229 important discoveries. Because the average output by age interval is greatest for the 30- to 34-year age group (being .45), their average output is set equal to 1, and the output for the other groups is expressed as a percentage of this. Thus the 25- to 29-year age group that produces .25 contributions per scientist alive in the age interval is reported to produce 55 percent (.25/.45) of what the most productive group contributes.

Almost without exception Lehman found that output among creative scientists for their best, or approximately best, work reached a peak during their thirties and then gradually declined. A typical profile is drawn in Figure 4-1 for 244 chemists who made a total of 993 significant contributions. Note that although Lehman's findings are consistent with an early peak and with the idea that the most likely age for eminent scientists to do exceptional work is in their thirties, the gradual decline with age means that a significant amount of work is produced by older persons. Indeed, Lehman was quick to point out that although the average output was highest when the scientist was young, in most of the fields he studied about half of all significant work was done after age 40.[4] Lehman also found some variation among fields. In physics and chemistry, for example, average production was highest in the early thirties, whereas in the field of astronomy, the peak came a good ten years later. In most other fields, the peak occurred in the late thirties.

In addition to examining the age at which eminent scientists made their most outstanding contributions, in a few instances Lehman extended his investigation to include lesser works of the eminent as well as "run-of-the-mine" works of the not-so-eminent.[5] Lehman found that these "quantity" profiles generally peaked later and were flatter than the quality profiles based on the significant contributions of the eminent.

After Lehman's work was published, particularly his book in 1953, it became fashionable to engage in what might be called "Lehman bashing."[6] The general argument was that Lehman's profiles were largely spurious because the methodology used weighted the odds toward the creative achievements of the young. Although many criticisms were voiced, they fall into four main catego-

Table 4-1. Hypothetical Example of Lehman's Methodology

Age interval	25–29	30–34	35–39	40–44	45–49	50–54	55–59	60–64	65–69
Number of scientists alive	100	100	100	98	96	93	89	82	73
Number of important contributions	25	45	40	35	30	23	18	10	3
Average output	.25	.45	.40	.36	.31	.25	.20	.12	.04
Normalized[a] output	.55	1.00	.89	.80	.69	.55	.44	.27	.09

[a]The average output of each age group relative to the average output of the most productive age group.

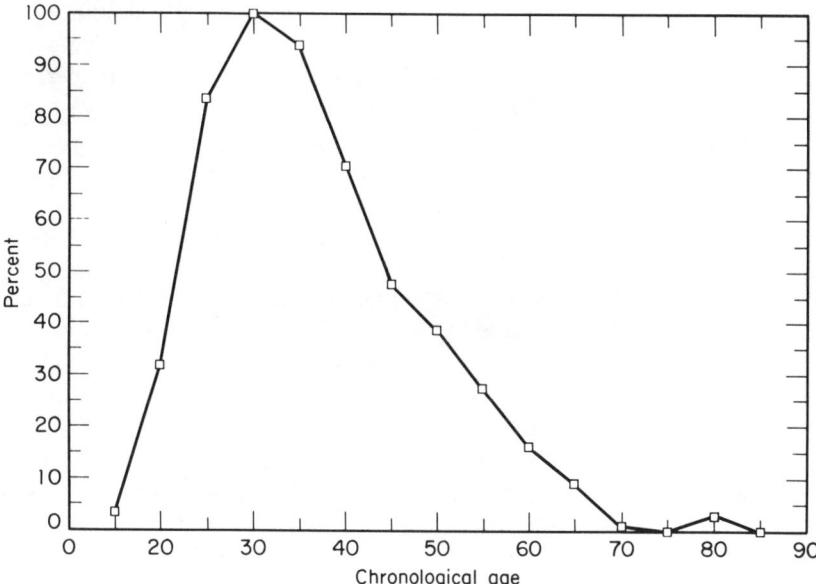

Figure 4-1. Age-publishing profile of 244 noted chemists making 933 significant contributions (the average output of each age as a percentage of the average output of the most productive age group). *Source:* Lehman (1953).

ries. Several focus on the rapid growth that science experienced for many decades. We discuss these here, as several critics were fairly successful in calling Lehman's work into serious question.

Until quite recently, science has been growing at an exponential rate. One consequence of this is that as a scientist ages, the competition for recognition increases. Wayne Dennis, among others, argued that this biased Lehman's work toward the young, in that younger members faced less competition.[7] Although this point is valid, the work of the eminent in Lehman's study was of such a high caliber as to, in all likelihood, be recognized regardless of the level of competition. Furthermore, the criticism makes sense only if a true longitudinal design had been used, that is, if Lehman had followed the same group of scientists over time in order to determine the ages at which their best works were done. For if this were the case, as members of the sample aged, increased competition in the field would lessen the likelihood that the best works of older scientists would be recognized. But Lehman used historical data. Hence, some of the young achievers in his sample were born in the late 1800s, others in the 1700s, and still others in the 1600s. It is difficult to claim, therefore, that age correlates with the intensity of competition in a field. Furthermore, Dean Keith Simonton has shown that in fields that actually shrank in size over time—and thus those in which competition presumably became less intense—the same type of age curve exists.[8]

Another criticism of Lehman's work is that because of growth, the population of scientists is at any time skewed toward the young.[9] Accordingly, even if the young are no more productive than older scientists are, at any point in time the young produce the lion's share of scientific output. Thus, if Lehman, as Stephen Cole[10] inferred, had used data collected from a single time period to examine the distribution of output by age and had sampled from lists of contributions rather than lists of contributors, a distribution skewed toward the young could be a statistical artifact. With rare exception, however, Lehman did not do this, but instead collected data for scientists over a long period of time. Only in the few instances in which he discussed quantity profiles[11] did he fall into this methodological trap.

A third criticism of Lehman's work focuses on the length of a scientist's life span, for, as one critic points out, "accomplishments by scientists who die young are necessarily made by young scientists. Had they survived, some of them probably would have produced additional important work at older ages as well."[12] Although this statement is obviously true, it is not relevant to the bulk of Lehman's work. For the most part, Lehman did not look at the distribution of output by age but, rather, at the average amount of output produced by scientists alive in each age interval. Consequently, the average cannot fall because of mortality. The only way that Lehman's methodology could bias the results toward the young is if scientists who died young were unusually productive compared with those who lived to a ripe old age. There is no evidence, however, that the exceptionally gifted die young. Indeed, Simonton points out that in her study of exceptional talent Catherine Cox found that only 11 percent of her sample failed to attain their fifty-ninth year, a full fifteen to twenty years after the expected peak for most scientific pursuits.[13]

A fourth major concern with Lehman's research was that historians of science may be prone to mention the earliest works of great men rather than the best works of great men.[14] If this were true, the results would be biased toward the young. Lehman contended that this was not the case, basing his argument on surveys of science historians' criteria for including works in their anthologies.[15]

Dennis's Work

There are several other studies of eminent scientists. In order to study the potential bias in Lehman's work pertaining to the life span issue, Wayne Dennis collected biographical information on eminent creators who lived through their seventy-ninth year.[16] Four fields of science were represented. The sample sizes were extremely small, with the largest field containing just forty-nine scientists. Rather than look only at the outstanding contributions of the eminent, Dennis measured total contributions, that is, the total number of works written each decade by the eminent scientists. Publication data were obtained from the *Royal Society of London's Catalog of Scientific Literature, 1800-1900*. Thus, Dennis studied only scientists whose careers fell entirely within the nineteenth century.

Overall, Dennis found that biologists' maximum output occurred in the decade of their forties, that their output fell somewhat during their fifties and sixties and was significantly lower in their seventies; botanists' output rose during their twenties and thirties and stayed fairly constant thereafter through their sixties; chemists' output peaked in their forties and fell off significantly after the age of 60; and geologists' peak occurred in their fifties, with their output declining thereafter.[17]

Dennis saw this as evidence that Lehman's work was biased toward the young. His conclusion, however, is a bit extreme, since Lehman's "quantity" profiles based on total contributions are generally flatter and peak later than do Lehman's "quality" profiles based simply on the best work of the eminent. Moreover, Dennis's work is not without its flaws. For example, the nineteenth century is often considered a century of "gentlemen scientists," and at least in fields such as botany and geology, it is quite possible that contributions remained high as these men aged, withdrawing from their other worlds to take up their scientific avocations more seriously. In addition, by restricting the sample to those whose careers spanned at least sixty years, entirely within the nineteenth century, Dennis, in essence, created several longitudinal samples of individuals born no earlier than 1780 and no later than 1840. It is entirely possible that their output increased with age as opportunities in science, especially the opportunity to publish, increased during the nineteenth century.

Nobel Laureates

The most obvious group of eminent scientists to study is Nobel laureates, for there is fairly wide consensus that in the three fields in science in which the prize is awarded it represents the ultimate accolade. Table 4-2 presents, for the period between 1901 and 1989, the distribution of laureates by field according to the age at which they did their prize-winning research.[18]

Several generalizations can be drawn from this table. First, the age distributions differ somewhat by field, a finding consistent with Lehman's research and the theoretical point of view discussed in Chapter 3 that to the extent age

Table 4-2. Percentage of Scientists in Different Age Groups When Doing Nobel Prize-winning Work,[a] 1901–89

Ages	Chemistry	Physics	Medicine	All fields
21–25	0.9	7.4	0.7	3.0
26–30	15.7	19.9	10.7	15.3
31–35	27.0	27.2	24.8	26.3
36–40	28.7	19.9	23.5	23.8
41–45	16.5	14.7	25.5	19.3
46–50	7.0	8.1	6.0	7.0
51–55	0.9	1.5	6.0	3.0
56–60	2.6	1.5	1.3	1.8
61–65	0.9	0.0	1.3	0.8
Mean age	37.6	35.7	39.0	37.5
Median age	36.5	34.5	38.0	36.5
Cases	(115)	(136)	(149)	(400)

[a]Age at the time the work was done, not published.

matters, it matters differentially across fields. The Nobel laureates in physics tend to be the youngest, and those in medicine or physiology (referred to hereafter simply as medicine) are usually the oldest. Second, although relatively few of the laureates are under 25 at the time they do their award-winning research, a significant number are in their late twenties, and even more are in their early thirties. Indeed, in physics, the highest proportion of laureates (about one quarter) do their work in their early thirties, and in chemistry, the highest proportion are in their late thirties. Furthermore, by the age of 40 almost three quarters of all the laureates in chemistry and physics have done their research, and in all three fields by age 45, over 85 percent of the laureates have done their prize-winning work. Third, the distributions, particularly in physics and chemistry, fall off precipitously after the age of 50. Indeed, for all three fields combined, less than 3 percent of all prize-winning researchers are over 55 at the time they do their work.

Because award-winning research often takes several years to complete,[19] the distributions in Table 4-2 were calculated using the "midpoint" age, that is, the age of the scientist at the midpoint of the research period. When the age distributions are recalculated in terms of the age at which the award-winning research was initiated, we find that over 80 percent of the chemists and fully two thirds of the physicists and laureates in medicine begin their prize-winning work before the age of 35. We also find that less than 8 percent of all laureates initiate their Nobel Prize-winning work after the age of 45, and only a handful (12 out of 400) after age 50. Looking at the data in this manner also helps explain the field differences noted earlier. In physics, for example, the Nobel Prize-winning work spans the shortest period of time, and it is this fact—not the fact that laureates in physics start earlier—that explains why laureates in physics tend to be the youngest. On the other hand, the difference between medicine and chemistry stems from the fact that the typical researcher in medicine (often a clinician) starts the award-winning research at a later age than do those in chemistry.

The numbers suggest that although it does not require extraordinary youth to do prize-winning work—as Dirac's fever chill verse quoted in Chapter 1 would have us believe—the odds do decrease markedly after the age of 50, and virtually no laureate undertakes prize-winning work after the age of 55.[20] In her work on Nobel laureates, Harriet Zuckerman presents evidence on the average (mean) age of laureates at the time they published—not conducted—the research that earned them the prize. (When more than one article was involved, the midpoint age was calculated.) Between 1901 and 1972, the average age for physicists was 36; for chemists, 39; and for laureates in medicine, 41.[21] From these averages, Zuckerman concludes that "this does not mean that the norm is a higher rate of significant discovery by youthful scientists. We may need to be reminded that reports of median ages at time of discovery tell us that half of the discoveries were made *after* the median as well as before."[22]

Although one must grant Zuckerman her point (at least if she had reported medians instead of means), an examination of the age distribution for all laureates over the period 1951 to 1972 reveals that the inference she drew is somewhat misleading. Yes, by definition, half of the laureates were older than the median age at the time they did their award-winning work. But when one asks how much older, one realizes that very, very few exceeded the median by more than ten years. Indeed, between 1951 and 1972, only about 10 percent of the laureates fell in the right-hand tail of the distribution. Einstein and Dirac clearly exaggerated the point: Extreme youth is not that great a blessing. But older age appears to be a curse.

"AVERAGE" SCIENTISTS

With few exceptions, most of the work concerning the relationship between age and scientific productivity in recent years has not been restricted to the eminent; instead, by sampling from a broad spectrum of the scientific community, these studies have focused on "average" scientists. In studying the age–productivity relationship for average scientists, several questions arise that are not an issue when only eminent scientists are considered. One question is how to measure scientific output. A second is the time period over which the data are collected. And a third is whom to include in the study. Before summarizing some of the studies of average scientists, we turn to a brief examination of these three questions.

Measuring Scientific Output

In a study of important contributions by eminent scientists, it is fairly straightforward to identify scientific output, since a specific or several specific inventions, innovations, or theories made the scientists worthy of a place in history. Thus, Charles Darwin is known for his work on natural selection, Alexander Fleming for his discovery of penicillin, Albert Einstein for his theories of special relativity and general relativity, Francis Crick and James Watson for their elucidation of the structure of DNA, and Barbara McClintock for her discovery of "jumping" genes. Identification is especially easy for scientists who have received eponymous recognition for their contributions to science: Avogadro's number, Planck's constant, Boyle's law, and Chargaff's ratios, to name just a few. When one studies the output of a sample of scientists drawn at random from the population of scientists, it is substantially more difficult to determine "what" and "how much" each scientist has produced. Contributions to science are not packaged in homogeneous units.

Three direct indicators of the amount of research produced during a period of time are generally available: counts of inventions, patents, and publications. Counts measuring the number of citations to a scientist's earlier work, although also readily available,[23] are not an indicator of the amount of research produced during a period of time. Instead, they serve as a proxy for the scientist's standing in the scientific community.[24]

Of the three measures, publication counts are the most widely used in studies of scientific productivity. And among publication measures, the number of journal articles is often chosen, as it is generally recognized that journals are the major outlet for recording scientific advances in many disciplines.[25] Moreover, the high acceptance rates of scientific journals—often in excess of 75 percent[26]—make publications a fairly strong indicator of whether a scientist is engaged in research. Furthermore, studies that compare bibliometric measures of scientific output, such as article counts, with other nonliterature measures of productivity, find high correlations, generally in the 0.6 to 0.8 range between the alternative measures.[27] Thus, there is a strong basis for using bibliometrics to study the production of scientific knowledge.

The number of publications, however, measures only the quantity of research activity with which a scientist is associated and not necessarily the contribution the scientist has made to knowledge, as publications obviously vary in quality and are often coauthored. For example, in particle physics, in which experiments often involve large teams, it is not unknown to have an author list that exceeds one hundred names, and it has been reported, perhaps apocryphally, that in at least one instance the author list was longer than the actual article.[28] Thus, in using publications as a measure of output, it is important to note whether the researcher has controlled for quality and attribution or has simply aggregated article counts, being content to combine major articles with articles of less merit and solo-authored articles with multiauthored contributions.

Collecting Data

Data on scientists are usually collected in one of two ways. *Cross-sectional* data are gathered at a moment in time, and *longitudinal* data follow people over a period of time. For example, the 1990 U.S. Census is cross sectional in design. Everyone in the sample was interviewed at approximately the same time. On the other hand, if the census were to choose a subset of families and reinterview them during the 1990s, the data would have a longitudinal component.

Both types of databases have methodological problems. In cross-sectional studies, age and generational (cohort) effects are intermingled, as we saw in Chapter 3 with the relationship between IQ and age. The reason is that in a cross-sectional study, the 60-year-old is not only 25 years older than the 35-year-old but was also born in a different era when values and opportunities may have been significantly different. Thus, if we find in a cross-sectional study that 60-year-olds engage in less research than do 35-year-olds, it is not possible to know whether the difference is related to age or is instead a consequence of what we refer to as cohort effects. (Recall that the decline in IQ with age found in cross-sectional data does not appear to be the result of aging but is instead the result of generational or cohort effects, because later cohorts, who are younger, have had access to better education.) Indeed, we believe the potential effects of cohort in science to be of such magnitude that we devote two chapters to their discussion.

This does not mean that using longitudinal data to follow a cohort over time is without problems. Consider a hypothetical study that interviewed the class of 1960 at graduation and then followed them over the next thirty years. Clearly, much more than age changes for them over time. For example, during this period, the probability of getting funding and being promoted changed. The importance of the double P's—publish or perish—increased. Thus, longitudinal data have the complication that the context in which work is done varies over time. Hence, longitudinal data, like cross-sectional data, are not without problems. Despite these limitations, with the exception of the work presented in Chapter 8, studies based on one or the other approach provide the best understanding to date of the age–productivity relationship and hence are summarized here.

Whose Productivity to Study?

The lion's share of scientific research is done at a select group of academic institutions. In certain fields such as physics and the life sciences, national laboratories and institutes also play an important role in research. In the earth sciences, a great deal of research occurs at the U.S. Geological Survey.

Most studies of the productivity of average scientists look exclusively at the academic sector not only because data are more readily available for scientists employed in academe but also because publications provide a ready measure of academic output. In other sectors, such as business and industry, it is harder to measure scientific productivity, since the incentive to publish and share knowledge is often discouraged by the organization's desire to establish property rights over the new knowledge and maintain secrecy until the subsequent profits can be captured.

Cross-sectional Studies

Two well-known studies of scientific productivity in academe were done in the 1970s using cross-sectional data.[29] One examined doctoral-level scientists employed at "prestigious" institutions, and the other looked at doctoral-level scientists across a broader spectrum of academic institutions. To be included in the former, it was necessary for the scientist to be employed in a Ph.D.-granting institution rated by the American Council on Education (ACE).[30] To be included in the latter, it was necessary to have taught at a university or college (either two- or four-year) that was included in the 1972-3 ACE survey of college and university faculty.[31]

For the prestige study, done by Stephen Cole,[32] random samples were drawn from scientists employed in 1970. The measure of output was the number of articles published between 1965 and 1969. No attempt was made to adjust these counts for the number of coauthors or the quality of the article. If one considers that on average these articles were 2.5 years old at the time the study was done and that there is about a two-year lag between the inception of work and publication date,[33] for any reported age category work began approximately four to five years earlier.

Cole reported the average number of articles by age interval for three fields of science: chemistry, geology, and physics. In chemistry, the average output was highest for scientists between the ages of 40 to 44 in 1970 (or more importantly, between the ages of 36 to 40 at the time the work was done) and lowest for the oldest category (those 60 plus), who published, on average, 63 percent of what the prime producers did.

In geology, Cole reported that output peaked for sample members in their early forties. Following our logic, this means that the work was done when the geologists were in their late thirties. In comparison with chemistry, there is a larger age decrement, with the output of the oldest group being equal to only 42 percent of that of the prime group. A similar story holds for physics: Output peaks for scientists in their early forties (or late thirties at the time the work was actually produced) and decreases after that, declining to 55 percent of the peak for the over-60 age group.

Alan Bayer and Jeffrey Dutton[34] studied the productivity of doctoral scientists in several fields teaching at a wide array of American colleges and universities in 1972–3. Although they constructed several indices of productivity, the measure that interests us here is the number of articles published in the preceding two years. Rather than relate publishing activity to chronological age, they related it to career age, the number of years since the doctorate was received. Furthermore, instead of simply reporting averages, they determined the "best-fit" model from a set of alternative linear and nonlinear forms for each field's age–publishing data.[35] Using this method, they found that output in physics increases rapidly for the first twelve years of the career and then drops very gradually, only to have another, lower peak toward the end of the career. The age implied at the high peak is fairly consistent with that found by Cole, since Cole's data encompass five years of publications, not two. The second and lower peak may be due to the selective retirement of less productive physicists toward the end of their careers.

Bayer and Dutton also examined publishing profiles of biochemists and earth scientists. For earth scientists, they found that article counts peak about ten years into the career (or, in terms of when they were actually written, somewhere in the late thirties), decline for a number of years, and then reach a second, lower peak toward the end of the career. In biochemistry, the pattern is substantially different: Article counts gradually increase until mid-career and then gradually wane.

It is also worth noting that unlike other studies, Bayer and Dutton reported the R-square statistic for each best-fit model. (R-square measures the percentage of the variation in the dependent variable—here, publishing activity—that can be explained by the age relationship specified. An R-square of 1 means that all the variation is explained, and an R-square of zero means that none of the variation is explained.) As Chapter 2 explains, productivity clearly depends on other factors besides age, factors such as the scientist's ability, motivation, and training as well as the availability of resources and colleagues. Although some of these factors may be related to the scientist's age (see Chapter 3), a host of random factors such as illness, campus politics, and the slowness of the reviewing process may also affect an individual's productivity. Even if one could adequately measure factors such as the scientist's ability or motivation, the presence of random occurrences means that at the individual level it is difficult to explain a high percentage of the variation in productivity. Moreover, when only one explanatory variable is used, such as age, one would be surprised if the best-fit model could explain more than 10 to 20 percent of the variation. The R-square values that Bayer and Dutton found, however, are generally much smaller than this, never being more than 6 percent, indicating that although a relationship exists, career age is a poor predictor of publishing productivity.

Little work has been done on the publishing productivity of scientists employed in nonacademic environments. One exception is Donald Pelz and Frank Andrews's study in the early 1960s of doctoral scientists employed in research and development laboratories.[36] They found two peaks or "humps" for published papers (as well as other measures of scientific performance). In development labs, the 45- to 49-year-old age group wrote the most papers during the previous five years; a substantial slump then ensues, followed by a "renaissance" ten years later. A similar pattern was observed in research labs, although here the slump and ensuing recovery occur five years earlier.

Pelz and Andrews also compared age–performance profiles across sectors: for research labs, university versus government; for development labs, industrial versus government. Within both lab settings, a two-peak profile is again revealed. Moreover, the peaks and troughs occur at the same age for the government and comparison groups of researchers, although in both settings the government slump is more pronounced and the renaissance is less dramatic than for the comparison group.

Longitudinal Studies

Given the expense of collecting longitudinal data, it is not surprising that few longitudinal studies of scientists have been done and that the ones that have been done are generally for small samples in specific fields. Indeed, we could find no examples of longitudinal samples in the physical or life sciences that were used to study the relationship between age and productivity.[37] Only studies of mathematicians are available, yet mathematics is so distinctly different from most of science that generalizations drawn from it are often suspect.[38]

Stephen Cole[39] collected data over a twenty-five-year period for a sample of mathematicians who received their Ph.D.'s between 1947 and 1950. For this group, Cole found minimal variation in average article production over time. On the other hand, the percentage who published dropped substantially over the period while the number of strong producers remained fairly constant. Because other things besides age changed during this period, such as journal space and the pressure to engage in research, it is difficult to know whether the results are related to aging or to other factors.

Arthur Diamond[40] used a slightly different methodology to study the publishing activity of mathematicians at the University of California at Berkeley between 1965 and 1975. Rather than follow a single cohort over time, Diamond found article counts for everyone employed in the department during each year of the study period and then "pooled" these observations and estimated a regression model. He found that the Berkeley mathematicians wrote less as they aged, even after controlling for the fact that during the later part of the decade mathematicians, in general, wrote more on average than they did earlier in the decade. It is, of course, possible that the older mathematicians in Diamond's sample published less because they came from cohorts that placed less value on publishing or because they suffered from knowledge obsolescence or discovered that their research interests had gone out of fashion.

The Productivity of Scientists Research (PSR) Study

In this chapter we have summarized work concerning the age–productivity relationship for eminent scientists as well as for average scientists drawn from the general population of scientists. For eminent scientists a fairly good case can be made that productivity is age related, particularly if one focuses on important work of the eminent.

It is more difficult to draw conclusions about the age–productivity relationship for the average science drawn from the general population of scientists. Two issues create particular problems: First, once one falls below the gods and returns to the playing field of mere mortals, the heterogeneity in scientific output increases greatly. Thus, it becomes significantly more important to have some means of controlling for the overall quality of research. Second, the question arises as to the proper time frame for data analysis. If cross-section data are used, the age relationship obtained may be the result of cohort effects, since older scientists in the study entered science earlier than did the younger scientists in the study. On the other hand, if longitudinal data are used to eliminate cohort effects, changes over time, presumably owing to age, could instead be due to changes in such things as competition and values (what we call time-period effects in Chapter 8) that occur during the study period.

In the early 1980s, with a view toward many of these problems and with the support of several foundations, we began a study of the productivity of "average" scientists.[41] The remaining portion of this chapter and all of Chapter 8 concern that research. In this chapter we briefly describe the database assembled for the study and how we measured scientific output. We then present basic age-publishing profiles. In Chapter 8, we employ a more sophisticated approach, one that controls for cohort and time-period effects, among other things, to create age-publishing profiles.

The PSR Data

Since 1973 the research branch of the National Academy of Sciences, known as the National Research Council (NRC), has been collecting data biennially for a sample of scientists who received their Ph.D.'s during the past forty-two years. Funding for this massive data-collecting project has come from the National Science Foundation. The data collected are referred to as the Survey of Doctorate Recipients (SDR).

The SDR is the largest and most comprehensive longitudinal study of scientists in the United States. For the purposes of studying scientific productivity, however, it has the fatal flaw of having no measure of scientific productivity.[42] Thus, a necessary component of our research has been to develop quantitative measures of scientific output that can be linked to the database. This was done first by matching scientists in the SDR with their publication record taken from the *Science Citation Index (SCI)*.[43] As we noted earlier, because there is about a two-year lag between research and publication, we matched scientists surveyed in a given year with publications written by them that appeared in scientific journals in the following two years.[44]

In addition to counting the number of journal articles that each scientist wrote during the two-year period, we also counted the number of authors associated with each article and constructed an "adjusted" — for coauthorship — count.[45] We also constructed a proxy for the quality of each article. The correction for coauthorship was straightforward, apportioning to each author a share of the article. Thus, if five authors were associated with one article, each author's share was counted as 0.2; if there was only one author, the article was counted as 1.0.

The proxy for quality requires a somewhat lengthier explanation. In addition to listing articles published, the *SCI* also keeps track of citations to each article. Although citations indicate a variety of things — such as the reputation of the author; the size, nature, and growth rate of the author's field; and the extent of cronyism in the field[46] — there is a strong consensus that citations also represent a measure of the value or quality of a scientific contribution. Heavily cited articles are generally better and have made a more significant contribution to science than have less heavily cited articles. It is not, however, practical when studying scientists to wait for an article's citation history to evolve, since this could take a considerable period of time. An alternative and more expedient way to make inferences about the quality of an article is to evaluate the quality of the journal in which the article appears and then to assume that if the article is in a top-notch journal, it, too, must be first-rate. In our study we use this approach. To control for overall journal quality, we use the "impact factor" developed by Eugene Garfield, the father of the *SCI*. This measure evaluates journal quality by determining how many times a journal has been cited relative to the number of articles published in the journal during some earlier time period, the presumption being that the scientific community reveals the relative worth of a journal's articles by more frequently citing those of higher quality. Thus, in our database a single-authored article appearing in a journal that has an impact factor of 6 is counted as a 6, and an article appearing in a journal with an impact factor of 0.4 is counted as 0.4.

These adjustments for coauthorship and quality mean that for each scientist we have four ways of counting the number of articles written during a two-year interval. PUB1, the crudest measure, counts all articles equally, regardless of the number of authors or the quality of the journal and is similar (except for the number of years over which it is counted) to the measure used by Cole and

Bayer and Dutton. PUB2 makes the adjustment for coauthorship, and PUB3 makes the adjustment for journal impact and hence quality. Finally, PUB4 adjusts for both journal quality and coauthorship.

Because of the great expense of linking the databases, we were forced to limit our study to four fields: biochemistry, a field at the forefront of the DNA puzzle and the ensuing genetic engineering revolution; earth science, a field encompassing great diversity and one in which the plate tectonics revolution took place during the mid-to-late 1960s; plant and animal physiology, a traditional or more "classical" field; and physics, a field widely studied by sociologists and one that figures prominently in the idea that science is a young person's game. Scientists are classified as being in a specific field if they received their doctorates in the field.[47]

Age-publishing profiles are given in Tables 4-3 through 4-6 by field and sector of employment for the full-time scientists in our study. In physics and earth science, the four publication measures are used; in the life sciences, adjustments were not made for journal quality, owing to severe funding limitations. The profiles show the average productivity of a group, normalized in the same manner that Lehman employed, so that an age interval's output is expressed as a proportion of the output of the most productive age group. The sectors chosen are "all sectors combined" (all scientists in the field employed in science or engineering), academia (colleges and universities as well as medical schools), FFRDCs (federally funded research and development centers, in physics), government (in biochemistry, earth science, and physiology), and business and industry. Furthermore, for physics and earth science the academic and the business and industry sectors include only those scientists working in their fields of training at the time of the SDR surveys. No such restriction was imposed in biochemistry and physiology because of the many interdependencies among the biological, chemical, and medical sciences. Although in business and industry, as we noted earlier, one expects a good deal of research never to be published because of its commercial nature and the need for secrecy, it is interesting nevertheless to see whether there is an age-publishing relationship in this sector.

The data used for these profiles combine four biennial SDR surveys. Each survey is linked with publications written by scientists in the two-year period immediately following the survey. The earliest cross section was collected in 1973 and uses articles published in 1974 and 1975 to measure output; the latest cross section was collected in 1979 and uses articles published in 1980 and 1981 to measure output. Consequently, age does not correlate perfectly with cohort (as it does in the two cross-sectional studies discussed earlier), since scientists of any given age could be born in one of six years. For example, in the combined database we could have two 30-year-olds. The one from the 1973 survey would have been born in 1943, and the one from the 1979 survey would have been born in 1949. Although the age-publishing profiles that we present are a simple and direct way to analyze the data, as will be seen in Chapter 8, we do not think that this is the best way to find "true" age-publishing profiles.

Table 4-3. Normalized[a] Age-publishing Profiles of Physicists by Output Measure

Output measure	<30	30–34	35–39	40–44	45–49	50–54	55–59	>60	Peak output
All sectors combined (7113 observations)									
PUB1	1.00	.98	.82	.79[b]	.78[b]	.69[b]	.49[b]	.34[b]	2.20
PUB2	1.00	.88	.77[b]	.73[b]	.68[b]	.60[b]	.51[b]	.38[b]	1.03
PUB3	1.00	.95	.77[b]	.73[b]	.69[b]	.65[b]	.39[b]	.28[b]	5.84
PUB4	1.00	.84	.69[b]	.64[b]	.56[b]	.53[b]	.40[b]	.30[b]	2.67
Academia (2525 observations)									
PUB1	1.00	.89	.67[b]	.70[b]	.68[b]	.57[b]	.48[b]	.31[b]	2.98
PUB2	1.00	.80	.62[b]	.62[b]	.56[b]	.42[b]	.46[b]	.30[b]	1.44
PUB3	1.00	.82	.58[b]	.58[b]	.59[b]	.59[b]	.37[b]	.23[b]	9.11
PUB4	1.00	.68[b]	.49[b]	.46[b]	.41[b]	.38[b]	.33[b]	.17[b]	4.59
Business and industry (889 observations)									
PUB1	.98	1.00	.79[b]	.60[b]	.42[b]	.40[b]	.58[b]	.11[b]	2.44
PUB2	.94	1.00	.85	.61[b]	.42[b]	.52[b]	.74	.15[b]	1.02
PUB3	.98	1.00	.71[b]	.55[b]	.34[b]	.39[b]	.46[b]	.07[b]	5.95
PUB4	.86	1.00	.74[b]	.54[b]	.35[b]	.51[b]	.61	.13[b]	2.49
FFRDCs (1196 observations)									
PUB1	.91	1.00	.75[b]	.54[b]	.65[b]	.71[b]	.34[b]	.24[b]	3.13
PUB2	1.00	.90	.67[b]	.48[b]	.55[b]	.63[b]	.29[b]	.28[b]	1.40
PUB3	.86	1.00	.72[b]	.51[b]	.52[b]	.52[b]	.26[b]	.19[b]	7.91
PUB4	.95	1.00	.68[b]	.46[b]	.49[b]	.49[b]	.23[b]	.30[b]	3.02

[a]The average output of each age group relative to the average output of the most productive age group.
[b]Significantly lower, at the .05 level, than the productivity of the most productive age group.

Table 4-4. Normalized[a] Age-publishing Profiles of Earth Scientists by Output Measure

Output measure	<30	30-34	35-39	40-44	45-49	50-54	55-59	>60	Peak output
All sectors combined (3555 observations)									
PUB1	1.00	.87	.86	.67[b]	.60[b]	.49[b]	.40[b]	.21[b]	1.56
PUB2	1.00	.73[b]	.67[b]	.54[b]	.48[b]	.45[b]	.38[b]	.21[b]	.99
PUB3	1.00	.85	.79	.63[b]	.57[b]	.40[b]	.29[b]	.16[b]	3.01
PUB4	1.00	.66[b]	.57[b]	.48[b]	.42[b]	.32[b]	.25[b]	.13[b]	2.02
Academia (1665 observations)									
PUB1	1.00	.89	.81	.64[b]	.57[b]	.44[b]	.39[b]	.19[b]	2.03
PUB2	1.00	.77	.68[b]	.55[b]	.48[b]	.37[b]	.36[b]	.19[b]	1.26
PUB3	1.00	.93	.78	.63	.55[b]	.37[b]	.27[b]	.14[b]	4.04
PUB4	1.00	.71	.58[b]	.49[b]	.40[b]	.26[b]	.20[b]	.11[b]	2.72
Business and industry (659 observations)									
PUB1	.89	.96	.46[b]	.47[b]	.85	1.00	.20[b]	.09[b]	.56
PUB2	1.00	.78	.41	.51	.71	.98	.25	.09	.35
PUB3	.60	1.00	.34[b]	.44[b]	.93	.98	.17[b]	.08[b]	.78
PUB4	.73	.77	.23[b]	.41[b]	1.00	.76	.19[b]	.05[b]	.51
Government (745 observations)									
PUB1	1.00	.87	.78	.77	.71	.63	.45	.32	1.33
PUB2	1.00	.67	.56	.52	.48	.67	.42	.29[b]	.89
PUB3	1.00	.65	.55	.69	.70	.43	.38	.28	2.64
PUB4	1.00	.56	.40	.52	.48	.53	.43	.28[b]	1.64

[a]The average output of each age group relative to the average output of the most productive age group.
[b]Significantly lower, at the .05 level, than the productivity of the most productive group.

Table 4-5. Normalized[a] Age-publishing Profiles of Biochemists by Output Measure

Output measure	<30	30–34	35–39	40–44	45–49	50–54	55–59	>60	Peak output
All sectors combined (4618 observations)									
PUB1	.77	.83[b]	1.00	.98	.83[b]	.69[b]	.65[b]	.57[b]	4.71
PUB2	.78	.86[b]	1.00	.99	.83[b]	.68[b]	.66[b]	.61[b]	1.82
Academia (2602 observations)									
PUB1	.65	.79[b]	1.00	.97	.93	.78[b]	.71[b]	.66[b]	5.57
PUB2	.75	.84[b]	1.00	.97	.94	.79[b]	.73[b]	.66[b]	2.18
Business and industry (974 observations)									
PUB1	1.00	.76	.72	.99	.39	.55	.44	.18[b]	2.38
PUB2	.87	.77	.75	1.00	.37[b]	.51[b]	.43[b]	.25[b]	.91
Government (440 observations)									
PUB1	.91	.94	1.00	.72[b]	.68[b]	.41[b]	.35[b]	.26[b]	6.36
PUB2	1.00	.96	.91	.75	.71	.40	.38	.35	2.30

[a]The average output of each age group relative to the average output of the most productive age group.
[b]Significantly lower, at the .05 level, than the productivity of the most productive group.

Table 4-6. Normalized[a] Age-publishing Profiles of Physiologists by Output Measure

Output measure	<30	30-34	35-39	40-44	45-49	50-54	55-59	>60	Peak output
All sectors combined (3214 observations)									
PUB1	.83	.91	1.00	1.00	.85	.70[b]	.90[b]	.45[b]	4.00
PUB2	.88	.95	1.00	.99	.84[b]	.66[b]	.67[b]	.56[b]	1.61
Academia (2291 observations)									
PUB1	.72	.83[b]	.94	1.00	.80[b]	.67[b]	.54[b]	.45[b]	4.82
PUB2	.81	.89	.96	1.00	.81[b]	.64[b]	.61[b]	.51[b]	1.92
Business and industry (381 observations)									
PUB1	.49	.74	.79	1.00	.92	.91	1.00	0[b]	1.65
PUB2	.36	.52	.62	.87	.50	.63	1.00	0[b]	.83
Government (303 observations)									
PUB1	.72	.98	1.00	.54[b]	.36[b]	.21[b]	.37[b]	.22[b]	3.77
PUB2	.85	1.00	.94	.50[b]	.38[b]	.21[b]	.38[b]	.36[b]	1.48

[a]The average output of each age group relative to the average output of the most productive age group.
[b]Significantly lower, at the .05 level, than the productivity of the most productive age group.

Age–Publishing Profiles in Physics

In addition to being faculty members, physicists work in a variety of other employment settings. Most important, these include FFRDCs (federally funded research and development centers), where 17 percent of all physicists were working in 1987, and business and industry, which employed an additional 35 percent of the Ph.D.'s trained in physics.[48] Important FFRDCs include the Fermi National Accelerator Laboratory outside Chicago, Brookhaven National Laboratory on Long Island, and the Stanford Linear Accelerator at Palo Alto, California. If the Superconducting Supercollider is ever completed, a significant number of physicists will also work at this research site. In business and industry, physicists are heavily concentrated in a few companies. In 1981, for example, nearly one third of the physicists employed in industry were found in just eighteen firms.[49] Companies employing substantial numbers of physicists include Bell Labs, Dupont, General Electric, IBM, RCA, Union Carbide, and Westinghouse.[50] Of the group, Bell Labs is by far the most prominent on the research front, having nurtured the work of several Nobel laureates, including John Bardeen, William Shockley, and Walter Brattain, who shared the 1956 prize in Physics for their discovery of the transistor effect.

Table 4-3 reports normalized publishing profiles for physicists. For each sector, four profiles are calculated, representing the four measures of output. The table (as well as Tables 4-4, 4-5, and 4-6) also shows whether the output of a specific age group is, statistically speaking, significantly lower than that found for the most productive group. The level of significance used for the test is 5 percent. (Loosely speaking, a 5 percent level of statistical significance means that there is a 95 percent chance that the results are not due to "luck.")

For physicists employed full time in science or engineering, the "all sectors combined" profiles show that in all cases the youngest group wrote the most articles and the oldest group wrote the least, and the least is substantially less. Putting it in absolute terms rather than in the relative terms of the table and using straight counts of articles (PUB1) as a measure, those under 30 averaged 2.2 articles in a two-year period, and those over 60 averaged .75 articles in the same time period. In mid-career, however, between the ages of 35 and 50, article production does not vary greatly with age, although it is somewhat lower than the output of the youngest group.

A comparison of the four profiles provides a means of determining how coauthorship patterns and the quality of the journal in which the article appears vary with age. The fact that the PUB2 profile falls off more rapidly than the PUB1 profile does, for example, means that older physicists write with disproportionately more coauthors than do younger physicists, although the profiles converge at the very end of the career. Furthermore, the fact that the PUB3 quality profile falls faster implies that relatively speaking, older physicists in the sample publish in lower-quality journals.

The story for physicists working in academia is similar. Output is highest for the youngest group, lowest for the oldest group. (The under-30 group writes, on average, 2.98 articles per two-year period, and the over-60 group writes just .92 articles.) Between the ages of 35 to 49, output stays at a fairly constant level. Adjustment for the quality of the journal steepens the decline, as does adjustment for coauthorship, at least until the last ten years of the career. Apparently the older physicists in academia do research in larger teams and publish the results in journals of lower quality than do their younger colleagues.

The pattern for physicists working (as physicists)[51] in business and industry appears, not surprisingly, somewhat different from the previous two. Output is highest for the under-35 age group, who in straight-count terms write about 2.44 articles in a two-year period, and is lowest for the oldest group, but during their fifties there is a spurt in output. Furthermore, the increase cannot be explained by the fact that administrators put their names on the research papers of their juniors, as the adjustment for coauthorship makes the increase in the fifties even more pronounced. On the other hand, there is still some indication that older scientists publish in less prestigious journals relative to those of their younger colleagues.

The age-publishing pattern is not quite as clear-cut for physicists working at FFRDCs. Again, we find those under 35 writing the most (in absolute terms, 3.13 articles per two-year period) and those 55 or older writing the least. In the middle years of the physicists' careers, however, in their early forties, output dips considerably, only to recover—at least in terms of the straight count and the count adjusted for coauthorship—over the next ten years. The adjustment for coauthorship suggests that over most of the career, older scientists tend to write with relatively more coauthors than do their younger peers. The adjustment for quality also indicates that older physicists write in journals of lower quality.

Age-Publishing Profiles in Earth Science

It is fairly common for earth scientists to work outside the academic sector. Indeed, in 1987 about 26 percent of the Ph.D.'s trained in the earth, environmental, and marine sciences worked in business and industry, and 21 percent worked in the government. In business and industry the primary employers are the oil companies; in government, the primary employer is the U.S. Geological Survey. Table 4-4 presents profiles for these two sectors as well as for the academic sector and for "all" earth scientists.

When we put all earth scientists together, regardless of sector, we find that the publishing profiles decline continuously with age. The youngest group is the most productive, writing an average of 1.56 articles per two-year period, and the oldest is the least productive, writing only about one third of an article. If we look at productivity adjusted for the number of coauthors, we find that the drop is slightly steeper until age 55, indicating that over most of their careers earth scientists tend to write with disproportionately more coauthors as they age. The quality-adjusted profile shows that older earth scientists publish their papers in journals of lower prestige than do younger earth scientists, a pattern similar to what we found for physicists.

A somewhat similar story can be told for earth scientists working in academic departments. The youngest group is the most productive, writing an average of slightly more than two articles in a two-year period, and the oldest group is the least productive, writing about one third of an article. Furthermore, the data suggest that over the course of the career, output declines continuously with age. Again we find that at least until the end of the career, productivity adjusted for coauthorship declines more rapidly with age than do straight counts of articles. On the other hand, the PUB3 profile suggests that the quality–age relationship does not differ markedly from the quantity–age relationship, at least until the scientists' fifties.

Earth scientists working (as earth scientists) in business and industry write fewer articles compared with their peers in academia. Moreover, their age–publishing profiles are substantially different. Depending on the measure, productivity tends to be relatively high, if not the highest, for those under 35, falls over the next decade, rises again between the ages of 45 to 54, and then once again declines for those 55 and older. This pattern, although more pronounced, is somewhat similar to that seen for physicists in business and industry. It is also reminiscent of the pattern observed by Pelz and Andrews[52] for doctoral scientists employed in research and development labs.

Earth scientists working for the government are relatively productive, writing about 1.33 articles per two-year period. Regardless of output measure, the most productive group is under age 30, and the least productive is over age 60. In terms of straight counts of articles, research activity declines continuously with age. This is not the case, however, when adjustments for coauthorship and quality are made.

Age–Publishing Profiles in Biochemistry and Physiology

Biochemists are employed primarily at four-year colleges, universities, and medical schools (53 percent in 1987), although some work for business and industry (28 percent), notably large pharmaceutical companies, and some at government labs (8 percent) such as the National Institutes of Health. In recent years the number employed in the profit sector has increased, particularly as a result of the revolution in biotechnology.[53] Physiologists are more likely to work in academe than are biochemists and less likely to work in business and industry. In 1987, for example, 66 percent were employed at four-year colleges, universities, and medical schools, 17 percent in business and industry, and 8 percent in government.

The age–publishing pattern for biochemists is quite different from the pattern we have seen so far in the physical sciences, particularly for the entire group of doctoral scientists and those employed in academia. The pattern, however, is somewhat similar to that observed by Bayer and Dutton for biochemists.[54] For both samples, publishing productivity is highest in the 35 to 39 age interval and is in excess of 4.5 articles per two-year period, noticeably above the rates found in physics and earth science. Furthermore, research activity remains fairly high throughout the biochemists' careers. The snugness of the PUB1 and PUB2 profiles over most of the career suggests that coauthorship patterns do not vary much with age in this field.

The profiles for physiologists look similar to those for biochemists. Again, for all sectors combined as well as for academia, peak productivity occurs in the 35 to 39 (or 40 to 44) age interval, and productivity remains relatively high over the entire age distribution. On the other hand, the profiles for physiologists working in business and industry look quite different from the profiles for biochemists in business and industry (or, for that matter, physicists or earth scientists employed in business and industry). Not only is publishing productivity in business and industry lower, on average, in physiology than in biochemistry, but during the twenty-year span from ages 40 to 59, publishing productivity, at least in terms of straight counts, is exceedingly high.[55] Perhaps this is a reflection of the dynamism of the biotechnology revolution affecting biochemists differently than physiologists.

Summary of the PSR Results

The profiles presented in Tables 4-3 through 4-6 show that in most cases article production varies with age. In the physical sciences, for the most part, productivity is highest for the youngest age group, those under 30, and lowest for the oldest group, those over 60. In the life sciences, output generally peaks toward the middle of the career and wanes only moderately with age. In both physics and earth science, for which "quality" profiles were computed as well as profiles adjusted for coauthorship, we generally found that relative to their youngest peers, older scientists are likely to have more coauthors and publish their works in less prestigious journals.

Like Bayer and Dutton, we also computed R-squares for the equations used to generate these profiles, and like Bayer and Dutton we found these R-squares to be quite low, never being higher than 7 percent. Thus, although there does appear to be an age–publishing relationship, we must stress that in the overall scheme of things, age explains only a small amount of the variation. Age does matter, but clearly many other things matter a good deal more.

Finally, we must caution the reader that the profiles presented in these tables (as well as the studies summarized earlier) are extremely crude in regard to the way in which the data are used. In particular, as we noted earlier in this chapter, not only are cohort and aging effects intermingled, but time-period effects are combined as well. Thus, we do not know whether the relationship shown is related to age or instead to some other effect. In Chapter 8 we examine the data in a manner that gives us a clearer picture of aging effects. Before doing so, however, we shall turn to two other topics. First, in Chapter 5 we explore the consequences that an aging scientific community has for both science and society at large. Second, in Chapters 6 and 7 we look at the importance of place and time, what we call RPRT, more fully.

NOTES

1. Goethe, in referring to Napoleon, stated: "One must be young to do great things" (quoted in Nelson 1928, p. 303).

2. A great deal of Lehman's work was published in his book *Age and Achievement* (1953) (see also Lehman 1944, 1958, 1962a, 1962b, 1963).

3. In some instances the reference works listed great scientists and their associated contributions. But in other instances, the reference works listed only outstanding scientific contributions, and the scientist associated with each was then determined.

4. See Lehman 1956, p. 334.

5. Potential problems arise with Lehman's methodology when he turns to the study of contemporary scientists. The issue is whether or not he begins with a sample of contributions and matches them to scientists or begins with scientists and matches them to contributions. When Lehman does the former to study contemporary scientists, which he apparently did on occasion, he commits a serious methodological error, because in growing fields a disproportionate number of discoveries are likely to be made by younger scientists simply because there are a disproportionate number of younger scientists alive at any time. This does not mean that the average productivity of younger scientists is higher than that of older scientists (see S. Cole 1979). When, on the other hand, Lehman takes a sample of contemporary scientists and then computes their average output, he avoids this problem.

6. Lehman's major critic was Wayne Dennis (see, for example, Dennis 1954, 1956a, 1956b, 1958, 1966).

7. See Dennis 1958.

8. See Simonton 1988, p. 257.

9. See the earlier discussion in note 5.

10. S. Cole 1979.

11. In Lehman 1953, chap. 15, a few tables examine the percentage distribution of total output by age.

12. Reskin 1979, p. 190.

13. Simonton 1988, p. 258.

14. Dennis 1956a, p. 332.

15. See Lehman 1960.

16. Dennis 1966.

17. Dennis used a similar methodology in an earlier article (1956b), though he grouped all scientists together regardless of their field. For this aggregated sample, which included scientists from astronomy, chemistry, geology, and mathematics, as well as naturalists, physiologists, and physicists, Dennis found output to be fairly constant during the three decades from age 30 to age 60.

18. Note that Zuckerman (1977) and Manniche and Falk (1957) report for earlier periods the age at which the prize-winning research was published. Here we report the age at which the prize-winning research was conducted. When it occurred over a period of time, as is often the case, we chose the midpoint. The primary source of information was Wasson 1987. Additional information was obtained from various issues of *Science, Science News,* and *Scientific American.*

19. This is particularly true when awards are given for a "line of research" rather than for a single discovery.

20. Keep in mind, however, that the age distribution is, without doubt, biased toward the young, as the Nobel Prize is not awarded posthumously.

21. Because both Zuckerman (1977) and Manniche and Falk (1957) queued on the age at which the prize-winning research was published, not performed, the means they report are probably one to two years older than the means reported in Table 4-2, because of the lag between research and publication.

22. Zuckerman 1977, pp. 165–6.

23. At least since the inception of the *Science Citation Index* in 1963 such data have been available (see Garfield 1983).

24. In a study of the relation of age to productivity, the appropriate dependent variable is the flow of research output measured over a specific period of time, not the stock of output produced up until a point in time.

25. See Menard 1971.

26. See Hargens 1975, 1988; Miller and Serzan 1984.

27. See Narin (1976) for a survey of twenty-four studies in which bibliometric measures were compared with other nonliterature measures of research productivity.

28. The record for the number of authors on a published paper is apparently 193. See *Science* 241 (1988): 1437.

29. Bayer and Dutton 1977; S. Cole 1979. Several other studies exist, including, for example, Blackburn, Behymer, and Hall 1978; Fulton and Trow 1974; Gillmor 1984, 1987; Kyvik 1990; National Science Board 1977; Pelz and Andrews 1976. For a review of some of these studies, see Fox 1983; Reskin 1979; Stephan and Levin 1987.

30. See Roose and Andersen 1970.

31. Bayer and Dutton 1977.

32. S. Cole 1979.

33. Garvey, Lin, and Nelson 1970, p. 63.

34. Bayer and Dutton 1977.

35. The criteria that they employ to choose the best-fit model are largely based on selecting the functional form for the postulated age relationship that explains the most variation in the dependent variable (the number of journal articles published in the last two years).

36. Pelz and Andrews 1976.

37. Allison, Long, and Krauze (1982) tested the cumulative advantage hypothesis for biochemists and chemists. Specifically, they tried to determine whether inequality in publications increases as a cohort of scientists ages, and they found that it does.

38. There are a variety of reasons that mathematics is considered distinct from science. One reason is that mathematics as a discipline is characterized by anomie, "the general absence of opportunities to achieve recognition." This failure in the operation of the reward system of science, it is hypothesized, leads mathematicians to behave differently from other scientists (Hagstrom 1965, p. 228).

39. S. Cole 1979.

40. Diamond 1986a.

41. The research was supported by grants from the Exxon Education Foundation, the National Science Foundation (NSF Grant #SRS 8306947), and the Alfred P. Sloan Foundation (B1983-43).

42. In the 1983 survey, a measure of output was included. For no other year, however, does a measure exist.

43. Institute for Scientific Information (ISI), annually.

44. For a description of the match and a discussion of its reliability, see Stephan and Levin 1988.

45. The maximum number of secondary authors listed in the *Source Index* of the *SCI* is 9. This limitation creates a problem only if subfields or specialty areas are not carefully defined. If they are well defined, the incidence of more than nine coauthors should be concentrated in a few areas, and any measurement error resulting will not likely bias the parameter estimates within these areas.

46. See, for example, Garfield 1983, pp. 244-9.

47. The field of training is indicated at the time the doctorate was earned and recorded in the Doctorate Records File (DRF) or at the time the scientist was surveyed by the SDR.

48. These numbers come from SDR tabulations compiled for us. They include U.S. scientists trained in the field who were employed full time in science or engineering in the United States.

49. Porter and Czujko 1981.

50. National Research Council 1972, pp. 592-4.

51. In academia the profiles are for physicists working in physics departments. Similarly, in business and industry, we restricted our analysis to those trained as physicists and employed as physicists in business and industry. The "all sectors combined" profile, however, includes all persons trained in physics working in science or engineering, not exclusively those employed in physics. Comparable definitions are used for earth scientists.

52. Pelz and Andrews 1976.

53. In 1973, 19.8 percent of the doctorates in biochemistry were employed in business and industry, compared with 28.4 percent in 1987. These data are from SDR tabulations.

54. Bayer and Dutton 1977.

55. Note that because of small sample size there are relatively few cases in either tail of the age distribution.

5
Does the Age Structure of Science Matter?

In this chapter we examine the consequences that an aging scientific community has for science as well as for society at large. We are particularly interested in whether an older community retards the speed with which new ideas are integrated into scientific theory and practice and whether an older scientific community slows the rate at which the economy grows. We begin by looking at the relationship of the age structure to the success of the scientific enterprise. We close by examining the relationship between science and economic growth.

AGE STRUCTURE AND THE RATE OF SCIENTIFIC DISCOVERY

A graying scientific community affects the rate of scientific discovery both directly and indirectly. The direct effect, as we saw in Chapter 4, is that older scientists, particularly those capable of making major discoveries, appear to be less productive than younger scientists are. Consequently, an older scientific community has less science in it than a younger community would have. Indirectly, an older scientific community can depress the rate of scientific discovery by affecting the speed with which new ideas are integrated into scientific theory and practice. Because we have already discussed the direct effects, we focus here on the indirect consequences of a graying community.

Age and Resistance

There are several well-known cases in the history of science in which an older scientist strongly resisted the innovative work of a younger scientist. Some of the most notable cases are the lifelong resistance of Lord Kelvin to Rutherford's theory of the electronic composition of the atom,[1] Joseph Priestly's similarly lengthy resistance to oxygen theory, Louis Agassiz's resistance to evolutionary theory, and Sir Harold Jeffreys's resistance to the theory of continental drift.[2]

Indeed, although Jeffreys may have had good reason to reject continental drift in the 1920s, his resistance persisted for decades, and he died a disbeliever in 1989.[3]

The notion that older scientists are slow to adopt new ideas and may actually impede the progress of science by blocking innovative work of younger scientists has been articulated by several scientists. Darwin, in a concluding passage of *On the Origin of Species*, wrote:

> Although I am fully convinced of the truth of the views given in this volume . . . I by no means expect to convince experienced naturalists whose minds are stocked with a multitude of facts all viewed, during a long course of years, from a point of view directly opposite to mine. . . . But I look with confidence to the future—to young and rising naturalists, who will be able to view both sides of the question with impartiality.[4]

More recently, in his autobiography Max Planck stated that "a new scientific truth does not triumph by convincing its opponents and making them see the light, but rather because its opponents eventually die, and a new generation grows up that is familiar with it."[5] Clearly, if Planck's "principle" is true, the changing age composition of the scientific community indirectly affects the rate of scientific advance.

Faced with a new idea, scientists may react in a number of ways. At one extreme are those who resist to the point of actively blocking efforts to communicate the new idea to the scientific community. At the other extreme are those who enthusiastically accept the idea and incorporate it into their own work; an occasional scientist may even become an early proselytizer of the innovative idea. Somewhere in between are those who tolerate the new idea, neither incorporating it into their research nor blocking its dissemination but instead continuing to work on as before.

One interpretation of Planck's principle is that older scientists are less likely to adopt new ideas than are younger scientists, choosing instead to continue doing the research they have done in the past. At best, according to this interpretation, older scientists tolerate new ideas. A stronger interpretation of Planck's principle is that older scientists are more likely to resist new ideas than are younger scientists and to use their positions of eminence and authority to impede the process by which these innovative ideas coalesce into a new paradigm. In its most damaging form, this can include an editor's refusal to publish articles espousing a different point of view, the denial of research funds to a scientist wishing to pursue an alternative approach, or the refusal to put on a program a scientist holding a different point of view. Strong advocates of one theory can also refuse to teach a competing theory to their students, thereby passing on only their own beliefs about a subject to a new generation. More subtle forms of blockage occur when members of an established school of thought fail to cite the work of members of another (opposing) school, thereby minimizing the amount of attention received by the new approach.

In this chapter we focus on social, economic, and cognitive factors related to Planck's principle. We begin by looking at theoretical underpinnings and conclude by reviewing several empirical tests of the principle. In this chapter we are interested primarily in how the age structure of the scientific community affects its aggregate productivity, rather than in how the age of an individual scientist is related to his or her productivity (the subject of Chapters 3 and 4). Even so, by necessity much of our discussion will concern the individual. Furthermore, as we shall see, our examination of how age relates to the acceptance of new ideas ties in with Chapter 3's discussion of age and productivity, particularly age and creativity. We talk about this here rather than in Chapter 3 because (1) the issue of the acceptance of a new idea is interwoven with the issue of resistance and blockage and (2) empirical tests of Planck's principle often use as the dependent variable the adoption of a new research agenda rather than the blockage of an idea.

Why Might Planck Be Right?

There are a host of reasons that older scientists may be slower to accept new ideas than younger scientists are. Economists argue that the accumulation of knowledge in an established theory gives older scientists a comparative advantage in working to improve that theory rather than in choosing to start afresh and learn all that would be necessary in order to appreciate and use a new, competing theory. From an investment perspective, the gains from adopting a new theory decrease with age, while the costs entailed in making the change increase.

It is also common to argue that younger scientists have the type of knowledge that makes them predisposed to accept change. Not only may their knowledge be more up-to-date; they may also have a broader background, having recently come from graduate school in which they were required to take courses across a wide range of topics. By contrast, older scientists may possess less up-to-date knowledge (see the discussion in Chapter 6) and, due to the specialized nature of research, have a narrower research focus.

There are also philosophical and sociological reasons to expect older scientists to be especially committed to the prevailing theory and less quick than younger scientists to embrace a new one. From a philosophical point of view, the commitment to a theory or paradigm provides the foundation for the puzzle-solving activity of "normal science" and the steady accumulation of scientific knowledge.[6] By reason of experience, older scientists are more likely than younger scientists to be firmly committed to the prevailing paradigm; they are the ones "who have used it to account for things previously inexplicable, who have experienced the range of its power and the difficulties of subjecting it to test."[7] Thus, when confronted with theoretical or empirical challenges to the established theory, older scientists may stand fast in their belief that "the older paradigm will ultimately solve all its problems, [and] that nature can be shoved into the box the paradigm provides."[8]

Social factors may also come into play in making older scientists resistant to change. For example, older, more experienced scientists may hold out longer than younger scientists do before accepting new ideas because they have more to lose in terms of professional reputation if the old ideas are proved wrong. The new ideas may discredit their leadership positions in schools of thought now being challenged or undermine their positions of authority in professional organizations, perhaps even taint previously awarded honors. Furthermore, discoveries by younger scientists, often of lower professional standing, may be resisted by older scientists of higher standing because the persons in higher authority view the relevant area of science as "their" area to do with as they please. Thomas Huxley recounts having great difficulty getting a paper published because a "particular" person in great authority "has come to look upon the Natural World as his special preserve, and 'no poachers allowed.'"[9] (Huxley believed so firmly in the inability of older scientists to change their minds that he declared that scientists ought to be strangled on their sixtieth birthday "lest age should harden them against the reception of new truths, and make them into clogs upon progress, the worse, in proportion to the influence they had deservedly won."[10]) The fact that Mendel's study of pea plants played such a small role in genetics for so long may not be, as is often assumed, because important scientists were unaware of it (there is ample evidence that Mendel's monograph reached many professional societies, universities, and academies at home and abroad and that he corresponded with one of the distinguished botanists of his time) but instead may be because the scientists in authority chose to ignore or discount work produced by an unimportant monk from a small town in Austria.[11]

The social perspective is well illustrated by an anecdote that the British embryologist Lewis Wolpert tells. It was 1968. Wolpert was 39 and had just developed some new ideas concerning the cellular basis of pattern formation in embryological development that, according to him, suddenly "made sense of an enormous number of different embryological results." That summer, after coming to the United States and giving an invited lecture at the Woods Hole Laboratory, Wolpert was disturbed to find that his ideas were met, at best, with silence. At a reception following the lecture, a leading American embryologist even turned his back on Wolpert. Confused about his cool reception, he asked a colleague for an explanation. The colleague reportedly replied: "Well, Lewis, they're all saying, who in the hell do you think you are?" They did not want to be challenged. To quote Wolpert, they "didn't like being told that they had been thinking about the problem in the wrong way."[12]

The social perspective also underlines the fact that the eminent are often extremely committed to an idea precisely because the idea is what made them eminent in the first place. The idea may even carry their name. Consequently, they have more to lose if proved wrong as well as more control over resources (after all, it is they who founded the "school" of thought) than do the noneminent. A study of forty-two prominent geologists doing research on lunar rocks reports findings consistent with this view.[13] The study examined how the forty-two responded over the course of the Apollo lunar missions to information and rock samples brought back from the moon. It became clear over the course of the interviews that three of the forty-two were especially attached to their own views of the moon, so attached that often the other thirty-nine used strong words to describe the attachment. One respondent, for example, told the researcher:

> X is so committed to the idea that the moon is Q that you could literally take the moon apart piece by piece, ship it back to Earth, reassemble it in X's backyard and shove the whole thing . . . and X would still continue to believe that the moon is Q. X's belief in Q is unshakable. He refuses to listen to reason or to evidence. I no longer regard him as a scientist. He's so hopped up on the idea of Q that I think he's unbalanced.[14]

Even so, on the basis of their previous work, the three most obstinate geologists were also judged by their peers to be among the most outstanding scientists in the sample. They had, one could argue, good reasons to resist change.

It is not only that older scientists have a greater investment, both socially and economically, in an existing paradigm and less to gain from change. New paradigms may call for new ways of thinking, and older scientists may be unable to use the new paradigm to reconstruct the world of science in their minds as quickly and easily as younger scientists can. Thomas Kuhn likens the process of scientific reorientation called for by a new theory to "a change in visual gestalt: the marks on a paper that were first seen as a bird are now seen as an antelope or vice versa."[15] "Thus, the very wealth of knowledge and manipulative skill of the experienced scientist makes it harder for him to accept innovations."[16] Younger scientists, by contrast, are more able to see the antelope precisely because they are less encumbered with past ways of thinking.

Taken together, these economic, social, and philosophical reasons make a convincing case for the contention that younger scientists are more likely to accept new ideas and to incorporate new ideas into their own research agendas than are older scientists. Not only do the young have a "fresh point of view." The young also have a greater incentive than established scientists do to adopt the new research agenda. In new areas there is less competition, and success offers the possibility of exalted reputation and enhanced financial remuneration. The same economic, social, and philosophical reasons also make a convincing case for the argument that older scientists are more likely to resist new ideas and in some instances to block new lines of research.

Such conclusions, particularly with regard to the incorporation of new ideas into a research agenda, may be mistakenly drawn, however, since they ignore the question of risk. Adopting a new line of research can be a risky proposition, much like investing in a speculative stock. The research may simply not work out. The scientist can spend months working on a theory or experiment only to find that it is a blind alley. And even if the research is successful, if it is not in the mainstream, it may be ignored by other scientists, published at best in a journal where virtually no one will see it. Work of a nonfrontier nature, ditchdigging work, on the other hand, is generally safer. Although the payoffs are not nearly as great, the costs are substantially lower, since ditchdiggers work on problems for which readily attainable solutions appear likely. Furthermore, ditchdiggers often work in areas of research in which "poaching" is encouraged, since poaching enhances the reputations of those already in authority.

Trivial Pursuits

The sociologist Peter Messeri believes that young scientists are less equipped to assume the risk associated with innovative research than are middle-aged and older scientists.[17] Older scientists have accumulated enough success to allow them to withstand a failure, but younger scientists have not. "Scientists accumulate social and intellectual resources as they advance through their careers, which allow them to absorb professional costs of pursuing lines of inquiry that deviate from established knowledge and practice."[18] Not only do senior scientists have research histories that will permit a failure; senior scientists also have job tenure, which gives them protection for working in unproven areas. Indeed, one reason often given to justify academic tenure is that it gives professors the freedom to pursue their own research agendas. By contrast, young scientists must constantly keep in mind the tenure and promotion clock when choosing a line of research. Because tenure decisions are generally made in the sixth year and it is not unusual to have what is called a third-year review in which those not making sufficient progress are given terminal contracts, the pre-tenure clock has fewer minutes than the post-tenure clock does. "The fact remains," as one author says, "that a published but trivial article is often valued more highly in university circles than the unpublished groping for an understanding of a particularly subtle facet of nature."[19] In science, the crime of aiming too high (and, apparently, failing), is a crime punishable by loss of job.

Does Planck's Principle Hold?

Planck's principle is a question open to empirical investigation. Unfortunately, it is far easier to test the relationship between age and acceptance of new ideas than the relationship between age and resistance to new ideas, since acceptance can be traced through publications, whereas rejection and the attempt to block ideas are more difficult to detect. Thus, whether or not the scientist has accepted a new line of research, as reflected in the scientist's writings, is often chosen as the dependent variable in empirical studies of Planck's principle. Sometimes even a narrower definition of acceptance is used, measuring attitude toward new ideas in terms of whether the scientists have adopted the new approach in their own research.

Before presenting the evidence, however, it is important to point out that there is simply no indication that the time between the introduction of a new idea and its general adoption is remotely as long as would be required for a whole generation to vacate their positions of authority. Even in the case of Planck, it would appear that not more than ten years elapsed between his first unsuccessful attempts to gain recognition for his reformulation of the second law of thermodynamics and its "universal acceptance."[20]

In recent years, five studies have investigated, in a nonanecdotal fashion, the validity of Planck's principle. The focus of two of them is the plate tectonic revolution that occurred in the mid-to-late 1960s; another examines the chemical revolution that took place in the latter half of the eighteenth century; a fourth study looks at the introduction of Darwin's ideas on evolution in the nineteenth century; and still another describes the introduction of quantitative models into the study of economic history in the 1960s and 1970s.

The relationship between age and the acceptance of Darwin's ideas is of particular interest, since Darwin, as we noted earlier, firmly believed that he would have difficulty convincing his colleagues, whose minds were "stocked with a multitude of facts all viewed, during a long course of years, from a point of view directly opposite to mine."[21] In an article appearing in *Science*, David Hull, Peter Tessner, and Arthur Diamond examined the writings of sixty-seven British scientists who were 20 years of age or older in 1859 (when *On the Origin of Species* was published) and who lived at least until 1869.[22] They applied two tests to Planck's principle.

First, the researchers looked at whether or not age is related to acceptance. Acceptance is indicated by whether the scientist stated, in either public pronouncements or private correspondence, that he believed in the evolution of species. The researchers found that those who failed to accept Darwin's ideas were, on average, about ten years older than those who became converts. They then attempted, by relating years of delay in acceptance to the age of the scientist, to see whether age influenced the rate at which the "acceptors" were converted to the Darwinian view. They found no evidence that younger scientists were converted more quickly than older scientists were, as Planck's principle would suggest. This result, however, must be viewed somewhat skeptically: The methodology of this portion of the study is flawed, for the researchers examined the relationship only for those who changed their minds, omitting from the analysis those who never accepted Darwin's ideas in the first place.[23]

In further work, Diamond looked at whether Planck's principle was operative in the adoption of cliometrics, the use of quantitative models in the study of economic history.[24] Although the group studied are economists, not "hard" scientists, the results could imply something about the predisposition of older persons to change. In this study Diamond defines acceptance more narrowly than he did in his earlier study with Hull and Tessner. Using a late 1970s' sample of North American members of the Economic History Association, he identifies as acceptors those considered to be cliometricians by virtue of their names appearing on a mailing list for a major conference in the new specialty. Diamond finds that younger persons are more likely than older persons to have accepted the new approach, a finding consistent with Planck's principle. On the other hand, age explains less than 10 percent of the variation in the likelihood of acceptance. Age was found to have a similarly low predictive value in Diamond's earlier work with Hull and Tessner.

One of the most revolutionary ideas to have emerged in the past thirty years in science is the concept of a dynamic earth. The idea is related to the concept of continental drift, a theory first put forth in the early part of this century, primarily by Alfred Wegener, who pointed to the close match of the continental margins on both sides of the Atlantic Ocean. (If one takes a map of the west coast of Africa and matches it up with the east coast of South America, the match, even to the untrained eye, is remarkably close.)

John Stewart[25] studied the social factors related to the acceptance of the continental drift theory during the period 1910 to 1950. One reason that Stewart's work is of particular interest is that unlike Darwinism, or for that matter cliometrics, the theory of continental drift did not become widely accepted during the period studied. Indeed, it was not until the theory of plate tectonics provided a global explanation for continental drift and other geological phenomena in the late 1960s that continental drift began to enjoy widespread acceptance.

To be included in Stewart's sample, scientists must have written an article related to the theory of continental drift. Thus, the sample includes only scientists who took a published stand on the theory. Consequently, it (and most of the other studies reviewed here) ignores those who "tolerated" the perspective but pursued their own research interests. For this limited group, Stewart found that age alone was not a predictor of acceptance. On the other hand, he found strong support for the idea that the more published the scientist is, the lower that person's likelihood is of accepting continental drift. Thus, although age per se was not revealed to be directly related to acceptance, the findings are consistent with the view that scientists with a larger investment to protect are less likely to be persuaded by innovative concepts.

More recently, Peter Messeri looked at the relationship between age and the acceptance of plate tectonics between 1956 and 1974.[26] Messeri's study is interesting not only because he uses a more sophisticated methodology but also because he divides the analysis into distinct time periods corresponding to how widely accepted the mobilist theory of shifting plates—as plate tectonics is often called—was at the time. Messeri estimated the rate at which those not yet converted adopted the mobilist research agenda. This technique, known as a hazard model, is better suited for determining the effects of age (as well as other factors) on the rate of acceptance than are the techniques used in previous studies. Acceptance in the study is indicated by whether the scientist adopted a mobilist program of research. In all, ninety-seven scientists who published in the area were included in the sample.

Messeri's results are consistent with a social interpretation of acceptance. In the early stage of the plate tectonic revolution, the adoption of the mobilist perspective was positively related to age. During the middle stage, there was no age effect, and during the later stage the effect of age was decidedly negative, when a "group of comparatively senior and more distinguished scientists came to make up an increasing proportion of the rapidly diminishing number who continued to hold out against the mobilist programme."[27] Messeri's interpretation of these results is consistent with the theory of risk articulated earlier. During the early years of the revolution, accepting plate tectonics was risky; thus young scientists, hot on the pursuit of promotion and tenure, chose "to work on orthodox research problems with high probability of yielding professional recognition in the short run."[28] As plate tectonics became more widely accepted, however, it proved less risky to be an advocate of the theory, and by 1974, the holdouts were mostly older, eminent scientists.

This emphasis on risk as an intervening variable is consistent with the findings from two other studies. In his 1988 book, Hull reexamined the data used by Tessner, Diamond, and himself earlier and found that when the sample is divided into ten-year age categories, both the oldest (those over 49 in 1859) and the youngest (those under 30) were the least likely to have accepted Darwin's revolutionary view of evolution.[29] Furthermore, H. Gilman McCann, in a study of the chemical revolution, discovered that during the early years of the revolution, middle-aged men with close ties to Lavoisier were the most likely to adopt the oxygen paradigm. Only during the period of "major conversion and consolidation" did a negative relationship between age and acceptance emerge, indicating that the defenders of phlogiston, the competing paradigm, were older men.[30] In addition, McCann argues that the age structure of the scientific community helps explain the rate of acceptance. In France, a younger scientific community more quickly adopted the oxygen perspective than in England, where an older community of scientists "held back the pace of acceptance of oxygen theory among British scientists of all age strata."[31]

To sum up, empirical studies of Planck's principle for the most part confirm the hypothesis that older scientists are slower than their younger colleagues are to accept new ideas and that eminent older scientists are the most likely to resist. The operative factor in resistance, however, is not age per se but, rather, the various indices of professional experience and prestige correlated with age. On the other hand, young scientists, as Messeri, Hull, and McCann found, may also be less likely to embrace new ideas, particularly if they assess such a course as being particularly risky. Thus, the extent to which Planck's principle operates depends in part on the context of what is happening at the time.

Two factors can influence the risk of undertaking an innovative research agenda. One—the degree to which other scientists have already begun to work in the area—has already been noted. The other—the state of the job market for young scientists—has not. Here we argue that the research agenda chosen by a young scientist is determined by what is expected of the scientist in order not only to become employed but, more importantly, to keep that employment. When jobs are scarce, when there are long lines of postdoctorates waiting for a permanent position, employers can afford to be selective both when they hire and also when the scientist is reviewed for tenure. When, on the other hand, job opportunities are expanding and positions are remaining vacant, employers are likely to be less selective.

An important determinant of the number of job vacancies for scientists, particularly in academe, is the age structure of the profession. In periods when a field is disproportionately old, as it is today, there are very, very few vacancies in a department. Consequently, employers become much more selective, raising the standards for tenure. The crossbar on the tenure hurdle is thus set higher. On the other hand, when the field is disproportionately young, the crossbar is lowered.

Does a higher tenure crossbar mean that young scientists seeking tenure are more likely to work on the frontier, to pursue innovative lines of research? Possibly, but not likely. If they strike the mother lode (not only in terms of producing frontier work but also in terms of getting it accepted), they, of course, will get tenure. But choosing such a course is extremely risky, especially when backwater research can generate numerous articles and tenure-and-promotion committees appear to be especially enamored of quantity. (It is not unknown for assistant professors to number the articles on their résumés, thereby making it easier for their seniors to count their output.) Not only does backwater research generate articles; safe research is often more attractive to funding agencies than are long-shot, risky projects. And the amount of funding dollars generated—not just article counts—is extremely important to the tenure decision. The end result is that when vacancies are scarce (precisely because the scientific community has become older), younger scientists are encouraged to "aim low," to produce less creative work. Thus, at the very time that there are fewer young scientists in the scientific community, those that do gain admission are encouraged by the system to go into more established areas of research, trading the risk of challenging the frontier for the safety of what some would consider mediocre work.

It is not just that the young are encouraged to do "trivial" work. They may also find it hard to get funding for their research. Indeed, it is far easier to renew a grant than it is to get a grant for the first time. *Science* reports that researchers seeking to renew existing grants are successful about twice as often as are newcomers applying for a first-time award.[32]

As we shall see in the next chapter, the labor market for scientists in the United States has been characterized by wide swings. Beginning in the late 1960s and continuing through most of the 1970s, new Ph.D.'s had difficulty finding jobs in the research sector. Initially, this scarceness of job openings resulted from the decreases in federal research support that accompanied the Vietnam War. Several years later, academic labor markets suffered another jolt when fewer applicants began to apply to graduate school in response to fewer job opportunities and also to cuts in federal support for graduate students. Thus, the scientific community aged. Here we have argued that this aging has had the effect of making the community more resistant to change as well as encouraging younger members of the scientific community to follow a safe, less innovative research path. In the next chapter, we focus on other ways in which an individual scientist is affected by events taking place in science at the time the degree is awarded. Before doing that, however, we turn to the relationship between science and economic growth.

SCIENCE AND ECONOMIC GROWTH

Our work in Chapter 4 and the studies just reviewed make us believe that an older scientific community is less productive than it would be if it were younger. First, there is some evidence that the productivity of average scientists is related to age, although the relationship is certainly not extremely strong. Second, both the Nobel laureate data and Lehman's work suggest that scientists who make major breakthroughs are most likely to be in their thirties. To the extent this is true, the extremely small proportion of scientists in the United

States today under the age of 35 should send a fever chill throughout the scientific community. Third, the evidence presented in this chapter suggests that an older community may retard the speed with which new ideas are adopted. Finally, the increased competition occasioned by the older scientific community means that young scientists may do less path-breaking work than they would do in a younger scientific community. All of this leads us to believe that our aging community of scientists has less science in it than a younger community would have. Clearly this is of concern to scientists. Should it also be of concern to nonscientists? Does science affect the larger society?

It is easy to see how science and technology affect our standard of living when one thinks of inventions such as the transistor and the integrated circuit, as we did in Chapter 1. Other examples of obvious relationships include the laser and its widespread application in the fields of medicine and communications and the array of promising agricultural and medical applications spawned by the discovery of recombinant DNA, or gene splicing, as it is popularly called. Human gene therapy, which also appears close at hand, provides another example of the link between basic science and the quality of life. Indeed, the possibility of genetic engineering emerged only after the structure of DNA and the genetic code had been solved in the 1950s and 1960s. And some would argue that these advances in basic knowledge could not have taken place without the earlier revolution in quantum mechanics.[33]

Just how valuable is science, especially basic research, to the larger society? How strong are the links between scientific research and the quality of life? Are we investing enough or too much in science? Such questions have always been important. Today, however, they take on a larger meaning in the world of Big Science with its multibillion-dollar expenditures for the Human Genome Project (to sequence the entire human genetic code), the Superconducting Supercollider (to study elementary particles and forces), the space station, and the Hubble telescope. With a combined price tag of over $45 billion, or approximately $180 for every man, woman, and child in the United States, there is renewed interest in knowing whether science really pays for itself. By spending such large sums of money, are we as a country investing wisely in the future, or are we merely adding to our national glory?

In order to evaluate the effectiveness of science, or at least the economic effectiveness of science, we must recognize that science can affect the standard of living in two ways. First, and in a way that is most visible to consumers, science and technology can provide new products—products such as CD players, VCRs, pacemakers, transistor radios, and notebook computers. Second, and in a less visible way, science and technology can affect the standard of living by improving the efficiency of the use of the nation's resources. Science and technology can facilitate getting more from the same quantity of land, labor, and capital; science and technology can lower the cost of production. In the parlance of economists, the state of knowledge can affect a nation's production possibilities frontier, which determines the maximum amount of both consumer and producer goods that can be attained at any given time. Advances in science and technology can move the frontier outward through the introduction of new and improved inputs (new types of equipment, a better-educated work force) as well as the introduction of new processes by which resources can be more effectively transformed into outputs. The end result can be economic growth: The society's national income can expand.

Such are the theoretical underpinnings for a relationship between science and technology, and economic growth. Do the facts support the theory? Is economic growth related to factors occurring in science? Although studies by Frederic Scherer and others[34] have shown that industrial research and development (R & D) has a powerful impact on productivity growth, the underlying role of basic science has largely been ignored. One simple reason for this is that it is very difficult to quantify the contribution of basic science and it is always simpler to assume that what is measurable is more important than what is not.

Recently, however, four studies have made some progress toward determining how important basic research is to the long-term growth of the economy. These studies start with the basic premise that academic science underlies, albeit with long lags, most recent technical change and growth; they then attempt to estimate the actual relationship between the two. The approach differs significantly from that of earlier work, which measured the contribution of science and technology to growth as a residual, that is, the percentage of productivity growth that could not be explained by increments in the basic inputs of land, labor, and capital.

In the first study, which focused exclusively on agriculture, Robert Evenson and Yoav Kislev found a statistically significant positive relationship between publication counts on corn and wheat research and subsequent changes in the actual output of corn and wheat.[35] That is, other things being equal, academic research led to increased agricultural productivity.

In a second and more ambitious study, James Adams[36] extended this line of research by investigating the relationship between knowledge creation and productivity growth across ten scientific disciplines and eighteen manufacturing industries. The broad coverage of the study enabled Adams to take into consideration not only direct effects (what are called *own* knowledge effects) but also indirect or "spillover" effects, that is, the flow of scientific knowledge from one field to another. Adams estimated that multifactor productivity growth[37] averaged 0.8 percent per year between 1953 and 1980, of which a substantial proportion (about two thirds to three quarters) could be attributed to growth in basic knowledge.[38] Not surprisingly, the lags between the accumulation of theory and increases in manufacturing productivity are long. For own knowledge, the lag was roughly twenty years between the appearance of research in the academic community and its effect on productivity; for spillover knowledge, the lag was even longer, on the order of thirty years. On the other hand, for more applied research, such as in engineering and computer science, the gestation period was shorter, about ten years.

Adams also documented the fact that multifactor productivity growth has fallen sharply since 1966,[39] a fact that has not gone unnoticed by others. Indeed, there has been increasing concern that the United States is losing its competitive edge, especially since the U.S. economy, although growing more rapidly in the 1980s than in the 1970s, continues to grow at a slower pace than some of its major competitors in the world market, notably Germany and Japan. Although the onset of the productivity slump has been dated at various times by different investigators, there is virtually universal agreement that 1973 was the watershed year.[40] From 1948 to 1973, multifactor productivity averaged 1.7 percent growth per year in the nonfarm business sector, but from 1973 to 1988, the rate slowed to only 0.2 percent per year.[41] Consider what this means to society in terms of forgone output (and income). Had output per unit of real input continued to grow throughout the rest of the 1970s and 1980s at its post-World-War-II trend rate of 1.7 percent per year, (nonfarm) private-sector output in 1990 would have been 30 percent higher than it actually was. This means that in 1990 dollars, the increment in GNP would have been about one trillion, nearly enough in increased income to pay off the national debt in three years.

In a more recent study,[42] Adams investigated whether resources are allocated efficiently in science, that is, whether the return to science is approximately the same across different fields. (The implication is that if the returns are not equal, a simple reallocation of funding from areas of low return to areas of high return will increase productivity growth.) Although the analysis is very preliminary, Adams is finding empirical support for an efficient funding regime. Furthermore, Adams has found that the estimated rates of return to science, although varying widely owing to possible measurement error, average in the neighborhood of 70 to 80 percent per year. Given that the risk-free return on capital has hovered at just 2 to 3 percent in recent years, one must wonder whether society is not grossly misallocating its investment spending. Adam's analysis in this study, however, must be considered exploratory. He himself has concluded that "much more study and more extensive measurement will be required before confidence in the evidence is sufficient to warrant a judgment on science policy and perhaps even a change in science policy."[43]

In the fourth study, Edwin Mansfield examined the extent to which academic research done in the past fifteen years has been instrumental in the development of new products and processes.[44] Data for the study were collected by surveying seventy-six major firms in seven industries: chemical, drugs, electrical, information processing, instruments, metals, and oil. Mansfield concluded that if the academic research had not been performed, there would have been substantial delays in about 10 percent of the product and process innovations commercialized during the period 1975-85. An additional 14 percent of the innovations were commercialized only after they received "very substantial aid" from academic research. Moreover, using very conservative assumptions, Mansfield estimated that investing in academic research yielded an average annual rate of return of 28 percent. Thus, Mansfield's findings provide additional evidence of the importance of basic academic research to economic growth and development.

Despite the apparent importance of academic science to productivity growth, the relationship between trends in science and the productivity slump experienced by the United States and other countries has largely been ignored. Although there has been some speculation that the productivity slump was in part due to declining investment opportunities occasioned by slower advances in science and technology,[45] most of the discussion has centered on explanations involving declining labor quality, rising energy prices (OPEC), environmental protection regulations, the depletion of mineral resources, and various measurement errors. If, however, one believes the studies of Evenson and Kislev, Mansfield, and Adams, the rate of knowledge accumulation in science must also be considered a factor. Adams, in fact, does hypothesize that the productivity slump may in part be a direct outcome of the slowdown in basic science that occurred because of World War II.[46] His numbers bear him out. He estimates that declines in knowledge account for some 15 to 30 percent of the slowdown in the multifactor productivity growth he observed between the periods 1953 to 1966 and 1966 to 1980.

Thus, there is a second reason to be concerned about an aging scientific community. Not only does a community of scientists that has become older have less science in it than a younger community does, but the price is paid by the larger society in the form of slowed growth and forgone output for years to come.

NOTES

1. Barber 1962.
2. These and other examples are in Messeri 1988.
3. See "Two Who Never Joined The Revolution," *Science*, November 3, 1989, p. 575.
4. Quoted in Kuhn 1970, p. 151.
5. Planck 1949, pp. 33–34.
6. Kuhn 1970.
7. Hagstrom 1965, p. 283.
8. Kuhn 1970, pp. 151–2.
9. Barber 1962, p. 550.
10. Quoted in Hull, Tessner, and Diamond 1978, p. 718.
11. Barber 1962, p. 551.
12. Wolpert and Richards 1988, p. 8.
13. Mitroff 1974.
14. Ibid., p. 586.
15. Kuhn 1970, p. 85.
16. Hagstrom 1965, p. 284.
17. Messeri 1988.
18. Ibid., p. 95. One other factor, according to Messeri, gives senior scientists an edge in taking on risky problems. That is, senior scientists are generally working on a larger number of problems than are younger scientists, and they therefore are more likely to be able to hedge their bets if part of their research fails. Their portfolio, in other words, is more diversified.
19. Quoted in Reif and Strauss 1965, p. 301.
20. Messeri 1988, p. 93.
21. Darwin 1859/1966, pp. 481–2.
22. Hull, Tessner, and Diamond 1978.
23. Because they estimated the relationship between years of delay and age (as of 1859) only for scientists who changed their minds between 1859 and 1869, their conclusion that age made little difference in the rate of acceptance could be incorrect. In fact, this would happen if the scientists who adopted the Darwinian paradigm during this period were relatively young and the scientists

who held out and did not change their views were relatively old. To eliminate this cause of error, a hazard model should have been estimated.

24. Diamond 1980.
25. Stewart 1986.
26. Messeri 1988.
27. Ibid., p. 107.
28. Ibid., p. 95.
29. Hull 1988, p. 380.
30. McCann 1978, pp. 91-92.
31. Messeri 1988, p. 93.
32. Palca 1990.
33. See Gribbin 1987, for example.
34. See Scherer 1986, for example.
35. Evenson and Kislev 1975.
36. Adams 1990b.
37. Two measures of productivity are commonly used in growth studies. The first and simplest is real output per hour worked, which is called *labor productivity*. The second and more complicated measure is *total* or *multifactor productivity*, the real output per unit of input (based on an index of all inputs used).
38. Article counts are used to construct an indicator of the stock of basic knowledge in each scientific discipline.
39. From 1.1 percent per annum in the period 1953 to 1966 to just 0.5 percent per annum in the period 1966 to 1980.
40. Scherer 1986.
41. These numbers are from Cullison 1989. The estimates of productivity growth vary according to whether one measures labor productivity or total factor productivity, as well as according to what special adjustments are made to the data series used.
42. Adams 1990a.
43. Ibid., p. 30.
44. Mansfield 1991.
45. See Nordhaus 1982 and Scherer 1986, for example.
46. Adams 1990b.

6
The Importance of Place and Time

The microbiologist James Watson studied with Salvador Luria at Indiana University, receiving his Ph.D. in 1950 at a time when DNA fever was in the air. Luria, renowned in his own right,[1] introduced Watson to the work of the "phage group"[2] and steered him toward the new frontier emerging in microbiology. It was also thanks to Luria that within a year of his graduation (at the tender age of 23), Watson was at work at the distinguished Cavendish Laboratory in Cambridge, England, headed by Sir Lawrence Bragg, himself a Nobel laureate. There he shared an office with Francis Crick. In less than two years they solved the structure of DNA.

The pharmacologist Candace Pert was a young (27-year-old) graduate student in the lab of "wunderkind" Solomon Snyder,[3] himself a protégé of the Nobel laureate Julius Axelrod, when early in 1973 she and Snyder discovered the opiate receptor—sites within the human brain that recognize opiates like heroin. Candace Pert had traveled a circuitous path to Snyder's door. Pregnant at 19 and married soon after, she followed her husband from Long Island to Philadelphia, where he enrolled in graduate school. While she was working as a cocktail waitress, a chance encounter with the assistant dean of admissions at Bryn Mawr led her to enroll at that school. Upon graduation, she applied to Johns Hopkins and the University of Delaware, the only graduate schools within commuting distance from where her husband was fulfilling his military commitment. Delaware accepted her; Johns Hopkins did not. Soon after, at a Bryn Mawr party, she met a Johns Hopkins professor of behavioral biology who urged her to write directly to "this weird guy Snyder," who was starting a graduate program at Hopkins, one quite different from the one that turned her down.[4] She did, and the rest is history. Three years later Pert and Snyder's breakthrough heralded the arrival of the new science of molecular psychology,[5] a search for the molecules of the mind that cause addiction, pain, emotions, depression, and mental illness.

James Watson, Candace Pert, and Robert Noyce (see Chapter 1) had the good fortune to be at the right place at the right time. They had RPRT. But it is not only Watson, Pert, and Noyce who had RPRT. Other scientists educated at approximately the same time in these emerging fields also had RPRT, although for them the effects were often less striking.

WHAT IS RPRT?

RPRT is a convenient way of saying that the careers of scientists are affected by events occurring at the time they receive their training. Two issues are involved. One focuses on what is occurring in scientific theory and practice when the scientist is being trained. The other focuses on whether the scientist can find a job in the top research sector after his or her training is completed.

Fundamental to the concept of RPRT is the idea that scientists' productivity depends not only on innate ability and curiosity but also on what they know and where they work. *Job market conditions* at the time scientists enter the professional labor market have a great deal to do with where they work. The *intellectual climate* in the field at the time of training affects what they know. In both instances, events somewhat out of the control of the individual influence the scientist's career path. Some scientists complete their training when jobs in the research sector abound; others, despite great intellectual promise, have difficulty locating a job in the top sector. Likewise, some scientists are educated when a revolution is occurring in their field, and they have the good fortune to become part of the revolution. Others are educated before a major innovation in theory or technique. All too early in their careers, they may find their research relegated to the sidewaters or backwaters of their fields. In this chapter we elaborate on the importance of RPRT.

JOB MARKET CONDITIONS

The Importance of Location

One reason that RPRT is so important is that scientists' productivity is related to the sector in which they are employed. Compared with their peers, scientists working at academic institutions and research centers are exceptionally productive, producing a disproportionate amount of basic research. Furthermore, within these sectors not all scientists are equally productive. Location in prestigious departments or centers such as the Fermi National Accelerator Laboratory in Chicago, one of a select group of federally funded research and development centers (FFRDCs), appears to foster individual research productivity.[6] For example, using publications written over a two-year period as a measure of research activity, we found that physicists employed at Ph.D.-granting universities wrote approximately three times as many articles as did those at other universities and four-year colleges, and twice as many as those employed in business and industry. At FFRDCs, publishing activity was also strong, although not quite up to the level produced at research universities. In the earth sciences, we found somewhat similar patterns. In the life sciences, also, scientists employed in graduate departments published more than did those working in nongraduate departments or in business and industry, although the differences are not as pronounced.[7]

Although the relationship between location and research productivity may be due in part to selective hiring, there is much to suggest that organizational context has its own effect, for it determines the character of research and how it is "conceived, managed, administered, and oriented."[8] Place makes a difference in science.

Several factors explain why location in prestigious departments and research centers fosters research productivity. First, these institutions provide better resources for research. One aspect of this is the quality of collegial exchange found at top research institutions.[9] Scientists working at such sites have lively colleagues and exceptional graduate students to work with and use for sounding boards. Another aspect is that the top research institutions facilitate the grantsmanship process. The name on the letterhead can be all-important in convincing an agency of a project's merit. Other aspects include lighter teaching or administrative loads and access to the most up-to-date equipment as well as to a supporting network of scientists devoted to keeping the facilities in top working condition. A second reason that place matters is that the acquisition of a prestigious appointment reinforces the scientist's values learned in graduate school concerning the importance of research. Scientists at top research institutions believe in research, and the appointment to a prestigious institution whets the scientist's appetite for recognition. Finally, place matters because it can jump-start the processes of cumulative advantage discussed in Chapter 3 and, through these processes, leverage previous productivity into even more productivity.

The location of a scientist within the scientific community depends a great deal on the job market conditions existing at the time the scientist completes his or her training. During some periods of time, jobs in the research sector are abundant; in other instances, not even Faust himself could get a job in a top research center. This means that certain scientists, upon completion of their graduate training, have disproportionate access to the resources that encourage productivity. Other scientists, such as those who completed their graduate studies in the 1930s during the Great Depression, are denied such access, not on their merits, but simply because they are at the wrong place at the wrong time.[10] Furthermore, and of particular importance to the RPRT perspective, these effects can persist over the entire career of the scientist, since upward mobility within the research sector is limited. Scientists relegated to the periphery for their first job may never gain access to the highly productive sector. On the other hand, scientists having the good fortune to get jobs in the research sector upon completing their graduate training can often remain in the productive sector throughout their careers.

In order to understand how market conditions relate to RPRT, we must look at how the market functions for jobs in science, particularly in the research sector. We begin by discussing the job market for new Ph.D.'s and then turn to the market for experienced scientists.

The Job Market for New Ph.D.'s at Research Institutions

The socialization process that commences during graduate school generally leads scientists to enter the job market with a strong preference to be employed at a Ph.D.-granting institution or a research institute. A study of physics postdoctorates found, for example, that following their postdoctorate positions, "most of these physicists were looking for tenure line university positions."[11] Another study found that industry was "not a career choice looked on with great favor in graduate schools."[12]

From the perspective of a research institution, the ready supply of applicants produced by the socialization process leads to a "buyer's" market. As a result, criteria must be established to make selections among the many candidates. One element in the decision is the prestige of the candidate's doctoral department. "One of the most persistent findings in the study of stratification in science is the substantial correlation between the prestige of the university department which currently employs a scientist and the prestige of his doctoral department."[13]

Research institutions, however, hire only when positions are available, and as we shall see, job market conditions for scientists fluctuate substantially over time. Indeed, the state of the labor market is the crucial determinant of whether a scientist's first job is in the research sector, since the preferences of scientists for jobs in the top sector do not appear to vary much by generation. In order to understand better why labor market conditions for entering scientists fluctuate so dramatically, it is useful to comment on properties of the supply and demand for scientists, especially the long adjustment process required in such a highly skilled market. After doing this, we shall resume our discussion, using one particular scientific job market cycle as an example.

Supply and Demand for Scientists

Historically, job openings for new Ph.D.'s in science have not unfolded at a steady rate but have varied substantially according to the level of resources available to the hiring unit. In academia, these resources depend on student enrollment and the amount of external funding, most of which comes from the federal government. Jobs at federally funded research and development centers (FFRDCs) are also heavily influenced by federal funding, whereas in business and industry the demand for scientists depends on the level of research and development spending (R & D), itself a function of the health of the economy.

On the supply side, the number of new doctorates also varies over time, partly in response to the job opportunities of recent graduates, which signal changes in the rewards for pursuing doctoral degrees in science rather than choosing other career options open to talented individuals. The pool of graduate students also depends on the size of the birth cohort from which the pool is drawn, as well as on the amount and type of graduate student financial support available. The supply can also be affected by the presence of what are viewed as unattractive alternatives, such as being drafted into the military service. Together, the interplay of supply and demand determines the state of the job market at any one time.

A feature that distinguishes the Ph.D. market from other markets is the long lag in the response of supply to a shift in demand. When consumers, for example, decided they wanted hula hoops in the 1950s (and again in the 1980s), it took no more than six months to supply their wishes. When the demand for research scientists increases, however, it takes years to meet the increased demand, since the production of a new Ph.D. in science, on average, takes over seven years.[14] Furthermore, given the extraordinary level of specialization, when the bottom drops out of the market, the highly trained supply does not instantly shift into another area. Because of such long delays in the adjustment process, employment cycles in scientific markets often last a number of years.[15]

From Boom to Bust: An Example of One Job Market Cycle

An Overview
From the late 1950s until the early 1970s, science in the United States experienced its longest period of sustained growth since the inception of doctoral education in the late 1800s. From 1959 to 1971, for example, there was approximately a threefold increase in the number of doctorates awarded annually in the sciences, a 2.5-fold increase in both the physical and earth sciences, and about a 3.3-fold increase in the life sciences.[16] These numbers translate into average annual growth rates of about 9 percent in all sciences combined, 7.5 percent in the physical and earth sciences, and 10 percent in the life sciences. Impressive in their own right, these numbers are even more impressive when compared with a long-run growth rate of just 6.6 percent in the number of Ph.D.'s awarded annually in all sciences.[17]

Much of this burgeoning growth in supply was a response to the increased demand for scientists caused by growing college enrollments and increased funding from the federal government for research. Toward the end of the period, the growth rate was reinforced as the baby-boom generation entered college in unprecedented numbers. Students during this period also were attracted to graduate education because of the prospects of a good job and because increased federal support in the form of research assistantships, fellowships, and traineeships made graduate education considerably less expensive.[18] In addition, as the Vietnam War accelerated in the late 1960s, graduate enrollment became a draft deferment strategy for some men.[19] Taken together, these conditions conspired to create unprecedented growth. Indeed, between 1966 and 1971, the number of Ph.D.'s granted in the sciences increased by almost 67 percent.

The Specifics
The period from the late-1950s through the mid-1960s is often described as a golden age for science in the United States. For many disciplines the launching of *Sputnik* ushered in a period in which jobs at top schools were plentiful. Federal funding was so widespread that academic scientists—parodying a pop-

ular scotch whisky advertisement of the time—were known to say, "While you're up [in Washington, DC], get me a grant." Indeed, from about 1959 until the Apollo space program was scaled back in 1967, federal support of academic R & D, which provided funding for research assistants and other personnel, as well as for research expenses and equipment, increased by almost 20 percent a year in inflation-adjusted dollars.[20] To meet the rising college enrollments and research demands, new universities were established, programs added, and faculties greatly expanded. Thus, for example, between the late 1950s and the early 1970s, the number of doctorate-granting institutions in the United States grew from 171 to 307; over the same period, in physics, the number of doctoral programs grew steadily from 112 to 194; in earth science, the number of programs more than doubled, from 59 to 121; and in the life sciences, the increase was from 122 to 224.[21] Overall, the full-time instructional staff employed in higher education grew from 154,000 in 1960 to 369,000 in 1970, an increase of almost 140 percent.[22]

In many disciplines the market for scientists then collapsed in the early 1970s. In some fields serious rumblings were felt as early as 1967. Particularly hard hit were scientists involved in the cold war–inspired physical sciences, in aerospace, and in energy-related fields. Growth in federal R & D slowed, and universities that had swelled their ranks with new assistant professors throughout the sixties now realized that they had enough and, in some fields, too many professors for the smaller birth cohorts expected to enter college in the late 1970s and 1980s. Indeed, the situation was so severe in physics that as early as 1970 the job market was described as "catastrophic."[23] The life sciences suffered less, partly because of growing federal support for research initiatives in cancer and heart disease and partly because of President Richard Nixon's "war on drugs."[24] They were also cushioned by the expansion of medical schools during this period.

One indicator of the change in job market conditions during this time is the increase in the percentage of new Ph.D.'s who indicated that they were seeking employment but had no specific job plans at the time of graduation.[25] In good times, most Ph.D.'s obtain a job (or a postdoctoral position) long before they defend their dissertations. In bad times, many scientists defend their dissertations having no idea where they will go after graduation. When the job-seeking statistics were first collected in the late 1950s, about one in twenty physical scientists who had planned employment had no specific job prospects at the time of graduation. Although there was a slight increase in the mid-1960s, the one-in-twenty figure was the approximate rule of thumb for the ten years between 1958 and 1967. Conditions deteriorated rapidly thereafter, with approximately one out of ten scientists reporting no specific job prospects in 1969, and close to one out of five, on average, reporting no specific job prospects over the period 1971 to 1977. By the late 1970s, conditions improved substantially, only to deteriorate once again in the 1980s.

In the earth sciences, labor market conditions were particularly strong in the mid-1960s, with fewer than 5 percent of the new Ph.D.'s who had planned employment reporting no employment plans at the time of graduation. By

1973, over 14 percent were still seeking employment at the time they received their degrees. Conditions remained generally depressed throughout the 1970s but improved in the early 1980s.

Although market conditions were significantly better in the life sciences during this period, they were not immune to market forces. During much of the 1970s as many as one out of every seven new Ph.D.'s in the area who wanted a job had not yet lined one up at the time of graduation. The major difference between the life sciences and the other sciences, however, is that a significantly higher percentage of people seeking employment in the research sector were eventually successful in locating a job. It may have taken a while, but life scientists were more likely than other scientists were to find research-oriented jobs during this period.

One way in which the scientific labor market coped with the excess supply of scientists was to create more postdoctoral positions. Although such positions provide valuable training, they also serve as a holding tank for talent yet to be harvested. Before the 1960s, taking a postdoctoral appointment was unusual in science. But by the mid-1970s it had become, if not the norm, extremely common in the physical and life sciences. In the physical sciences, for example, about 23 percent of new Ph.D.'s in 1968 took a postdoctoral appointment after graduation. By 1978, the number had increased to 40 percent, and by 1988 it was 48 percent. A somewhat similar situation occurred in the life sciences, in which postdoctorates grew from 35 percent of new Ph.D.'s in 1968 to 61 percent in 1988.[26]

It is not only that scientists became more likely to take postdoctoral positions during this period. Postdoctoral positions also began to last longer. Indeed, in some instances, scientists who were unable to get a permanent job moved from one postdoctoral position to another. In the physical sciences, for example, in the late 1970s as many as one out of every ten Ph.D.'s who had been out for four years had a postdoctorate, and by the late 1980s this number had climbed to one out of every eight.

The picture sketched by the job-seeking and postdoctorate statistics is one of general malaise in the market for scientists, especially physical scientists, during the late 1960s and 1970s. Although because of the way the data were collected it is difficult, if not impossible, to calculate directly the employment rates for new Ph.D.'s in the research sector during this time period,[27] it is fairly clear that a whole generation of physical scientists, and to a much lesser extent life scientists, left graduate school during this time only to find the job of their dreams, the job for which they thought they were preparing, difficult, if not impossible, to obtain.[28]

The Determinants of Current Location

Labor market conditions existing at the time that doctoral training is completed affect not only scientists' initial placement but also their subsequent career moves. That is, the location of experienced scientists within the scientific community can largely be traced to successful placement at the time their doctorates or postdoctorates were completed, for, with the exception of superstars, upward mobility in academia is limited. Nor are scientists likely to move from industry or government into the top research sector unless they have established a national or international reputation or unless a shortage develops.[29] Such reputations, however, are difficult to build when the initial job is at a peripheral institution. And scientists working in the periphery have difficulty communicating with top scientists. They have less access to resources and are encumbered by heavy teaching loads. Their research activity is depressed, and this depressed productivity affects their future productivity by lowering their taste for recognition and short-circuiting them from the processes of cumulative advantage. Furthermore, the story does not end here. Five to ten years later, even if jobs in the top research institutions become more available, they are unlikely to move up in the hierarchy because they lack the reputation and publications needed to make the shift. The effects of RPRT thus can be felt throughout scientists' careers.[30]

One way to see the imprint of the market on scientists' careers is to look at how Ph.D. classes of scientists are distributed today across the various employment sectors. Table 6-1 does precisely this for the physical sciences, the area most hurt by the events that began to unfold in the late 1960s. Three employment sectors are featured: research universities and medical schools,[31] federally funded research and development centers (FFRDCs), and business and industry. Scientists employed elsewhere (such as in four-year colleges or government) are included in the "other" category. Scientists are placed into nine distinct groups. The earliest group encompasses the 1944-to-1958 Ph.D. years because the small sizes of these classes limit reliability if the data are more disaggregated. Each row gives the proportion of scientists in a particular group employed in the various sectors as of 1987. If all groups had had equal access to the top research sector, approximately the same proportion of each would have been employed in this sector in 1987.[32]

The table tells a story of unequal access. The declining percentages in the research university and medical school column and the increasing percentages in the business and industry sector are impressive. Consider, for example, the

Table 6-1. The 1987 Employment Distribution of Physical Scientists by Year of Ph.D. (by percentage)

Date of Ph.D.	Research universities and medical schools	FFRDC's	Business/ industry	Other
1944–58	33.9	9.5	36.2	20.4
1959–61	31.1	5.2	37.3	26.4
1962–64	27.0	10.6	34.4	28.0
1965–67	22.4	9.9	40.6	27.1
1968–70	17.3	7.8	44.3	30.6
1971–73	16.0	11.0	50.0	23.0
1974–76	17.1	12.0	56.9	14.0
1977–79	14.9	12.2	56.4	16.5
1980–82	17.3	7.3	60.5	14.9

Source: Survey of Doctorate Recipients (SDR) tabulations.

fact that one in three members of the two earliest groups is employed today in a research-oriented academic institution, whereas for members of the mid-1960s group, the figure is about one in five. The imprint of the late 1960s and early 1970s is even stronger. Only one of every six Ph.D.'s who graduated during this period is now working at a top research university. Moreover, the situation has not improved for more recent classes. Whereas fewer than four out of ten members of the earliest group work in business and industry today, six out of ten of the latest group do.[33]

The numbers are even more dramatic if one focuses only on what has taken place in physics. Although not reported in the table, there is about a fifty-fifty chance that a member of one of the earliest classes of physicists works today at a research university. The comparable probability for a member of the 1971-9 group is about 17 percent. The opposite is, of course, true for the likelihood of being employed in business and industry. Whereas only one out of five of the earliest group works today in business and industry, more than four out of ten of the most recent group are employed in that sector. Again, this is a story of unequal access.

Growth in the Field

The forces of RPRT depend in part on a field's growth characteristics. Most blessed is the scientist who enters a field in its earliest period of growth. Most cursed is the scientist who joins a field on the verge of stagnation. One reason that some life scientists survived the perils of the 1970s much better than did physical scientists is that their field was generally in a period of sustained growth during this time.

The oceanographer and historian of science, Henry W. Menard, eloquently states the case for growth: A scientist who enters a field in its dynamic phase

has his professional goals within his reach. As a student he receives support from the generous research grants of his professors. He publishes several papers and is an established professional about the time he receives his doctorate. He is showered with job offers from prestigious institutions seeking to expand into his specialty. These continue after he is employed, and his promotions accelerate. He is invited to speak at international conferences and hobnobs with the mighty.[34]

But the fate of the entrant into a much slower-growing field is likely to be very different:

His professors have little money for his support. He is fortunate if his thesis is issued as his first publication a few years after he receives his doctorate. Every department already has a man in his specialty, so openings are available only when retirements happen to occur. It is unlikely that he will receive many job offers, and he may have to abandon his hopes for research in his specialty in order to take a job teaching . . . in an undergraduate college. Promotions will be slow or average. If he can do research, perhaps on his own time in summers, he can look forward to long years before he rests easy among the masters of his subject.[35]

Thus, RPRT depends not only on the job situation at the time of graduation but also on the rate at which the field grows after graduation.

Many Americans' involvement with science began when the science train, if it had not just pulled out of the station, still had a long way to go before it began to slow down. This was especially true for those who received their Ph.D.'s before the late 1960s or early 1970s. Since that time, however, there has been very little growth in science. Especially hard hit have been the physical sciences. Thus one would expect that market conditions have been most severe for those physical scientists who received their doctorates in the late 1960s and early 1970s. Not only did they have an exceptionally difficult time getting a research-oriented job when they graduated or completed their postdoctorates; for them the market has remained relatively weak.

INTELLECTUAL CLIMATE

Theoretical particle physicists trained in the 1950s were taught (some would almost say indoctrinated into) field theory. Earth scientists who trained at Columbia University in the 1960s were taught that the oceans were permanent features of the earth. Biologists trained in the 1940s learned that proteins were the key to life. Mathematicians and physicists educated in the 1980s were exposed to chaos as a formal science. Biochemists trained in the 1950s learned to use radioactive chemical tracers to discover where a drug exerts its action in the body. Molecular biologists educated in the 1970s were taught the technique of recombinant DNA, or gene splicing. Experimental physicists in the 1940s and 1950s used lantern slide projectors to produce beams of light; in the 1930s and 1940s they used vacuum tubes in experimental apparatus. X-ray crystallography — a technique for studying the structure of molecules by shining X-rays through them — was an important component of a chemist's training in the 1940s. By the late 1980s, supercomputers had become an important tool in theoretical physics as well as in other areas requiring large-scale computations and model simulations.

These are examples of theories or techniques considered to be state of the art at the time they were taught. Some still are considered state of the art (chaos, use of radioactive chemical tracers, gene splicing, and x-ray crystallography) and remain an integral component of a scientist's training in the field. Others, such as the role of proteins in genetics, antidrift theory, the use of lantern slide projectors, and vacuum tubes, are not. They have been replaced by new theories or techniques: DNA in the case of proteins, plate tectonics in the case of antidrift theory, lasers instead of slide projectors, and transistors and integrated circuits instead of vacuum tubes. And parallel computers may someday replace supercomputers. Finally, one of them, field theory, was declared out of date in the late 1950s and 1960s only to resurface as an important theory in the 1970s.

The Concept of Vintage

The key to the intellectual climate argument is that scientists' productivity is related to what was happening to knowledge in their field at the time they were trained. An overwhelming characteristic of science today is the speed at which knowledge grows and changes. Change, however, is not constant across fields, nor is it constant over time. From time to time in science, changes occur that are fundamental enough to render previous thought in the area irrelevant or, at most, relevant only from the point of view of a historical perspective. Sometimes these changes are dramatic and far-reaching, as in the case of quantum mechanics, the concept of natural selection, the theory of plate tectonics, or the discovery that genetic information is encoded in DNA. Sometimes these changes are less dramatic, or at least appear to be less global in their implications, as in the case of many body theory,[36] models of the atomic nucleus, or approaches to molecular structure. Nevertheless, such changes can have a major impact on scientists working in the area even if their wider impact is smaller or the general public never learns of them.

Change in science is not limited to theory. Research techniques and protocols are also subject to change. Thus, for example, the advent of the computer ushered in a whole new approach to science. Not only did computers enable scientists to solve problems heretofore unsolvable, they also altered "the form of scientific theories, in that logically linked propositions and formal mathematical statements have been replaced by complex computer models."[37] Likewise, the laser, first developed in 1960, provided a new technique for a variety of procedures. It was not just an additional piece of equipment; the laser displaced other techniques that, by comparison, were clumsy. The introduction of the transistor provides another case in point. Before transistors, researchers used vacuum tubes in experimental apparatus. Because vacuum tubes are macroscopic and transistors are microscopic devices, the use of transistors required an entirely different mind-set.

Changes in the intellectual climate raise the possibility that some scientists are particularly lucky, learning theories or techniques while in graduate school that remain relevant for an extended period of time. But other scientists are not, receiving their training in theories and techniques that rapidly fade from importance. In our terminology, scientists trained when intellectual conditions are ripe belong to a good vintage. Particularly fortunate are scientists who receive their training at the time the change is actually occurring and thus get in on the ground floor of a new approach or school of thought. Scientists, on the other hand, who have the bad fortune while in school to learn techniques and theories that soon fade from importance belong to a poor vintage. They must cope with more change than must peers coming from a good vintage, and hence their productivity may suffer.

Clearly, the time that a scientist is trained is extremely important to determining vintage quality. Place can also matter. Knowledge, particularly in the early years of a revolution, is not uniformly distributed across educational institutions. Consequently, some scientists are more fortunate than others, receiving their education at a more favorable location within the scientific community. They get on board earlier. For example, earth scientists educated at Princeton were exposed to the concept of plate tectonics earlier than were earth scientists at other universities. And physicists, as we saw in Chapter 1, who studied at Grinnell College in the late 1940s, had a head start in learning solid-state electronics. In addition, some scientists have the opportunity to work with outstanding mentors who are not only geniuses at their craft but also have the knack of nurturing genius in others. Indeed, neuroscientist and opiate-receptor researcher Solomon Snyder claims that "the best way to predict who will make a discovery worthy of a Nobel Prize is simply to examine who trained them."[38]

The fact that the quality of a scientist's vintage depends on RPRT does not mean that scientists have no control over their cognitive resources. True, change can make knowledge acquired at an earlier time obsolete. But whether scientists from a poor vintage themselves become obsolete depends on whether they replace these obsolete ideas with new theories and techniques. If they do, they will stave off obsolescence. But the process of learning takes time away from research and requires resources. Productivity is threatened. Scientists, on the other hand, who do not keep up have more time for research, but the science they produce may be more for their own enjoyment than for the enlightenment of the scientific community. Thus, in regard to doing research, scientists from poor vintages face the distinct possibility of being damned if they do keep up or damned if they do not.

The Threat of Obsolescence

Scientists who belong to a poor vintage are not the only ones who must be concerned with obsolescence. The threat of obsolescence hangs like a specter over all scientists. Because change is the very essence of science, even those coming from good vintages must be concerned with future changes that could render their knowledge obsolete. Anxiety about obsolescence is an everyday part of the scientist's life. In her in-depth investigation of the world of high-energy physicists, Sharon Traweek observes: "In the course of a career a physicist learns the insignificance of the past, the fear of having too little time in the present, and anxiety about obsolescence in the face of a too rapidly advancing future."[39] She goes on to say: "The established physicists are afraid that they will not continue making significant contributions, that they and their work will become obsolete."[40] Richard Feynman, one of the most brilliant physicists of the twentieth century, recalled a time in his career when "I was not really quite up to things: I was always a little behind. Everybody seemed to be smart, and I didn't feel I was keeping up."[41]

In our own work we have found that the mere mention of the word *obsolescence* strikes a raw nerve in some scientists. For example, some refused to talk about the vintage concept with us because obsolescence struck them as contrary to science. One eminent scientist even recommended that we change our research focus, for, to paraphrase him, well-trained scientists know how to remain current. Some scientists go so far as to imply that their training as scientists protects them against obsolescence, or as one seasoned observer of science told us, their Ph.D. "ordination" protects them from the threat of obsolescence. Other scientists, however, like Feynman, acknowledge obsolescence as a problem and discuss how it has affected their lives. One prominent scientist told us that his entire research agenda in later life had been shaped by the fact that he had become, at least in his own mind, obsolete. Others talked about going into backwater research as a means of coping with the threat of obsolescence.

In the following discussion we develop the vintage concept more fully. We begin with the concepts of growth and change in scientific knowledge, taking care to distinguish between the two. In addition, we consider the presence of fads in science. We then relate this discussion to the concept of codification, an analytic construct developed by sociologists. We conclude with the strategies that scientists can use to cope with the threat of obsolescence.

Growth and Change in Scientific Knowledge

Scientific knowledge grows at an astounding rate. Derek de Solla Price estimated that the scientific literature doubles every ten to fifteen years and that over 80 percent of all scientists who have ever lived are alive today. (The comparable figure for the world's population is 5 percent.)[42] The speed with which science grows surprises even scientists. In his autobiography Francis Crick states:

When I started biological research in 1947 I had no suspicions that all the major questions that interested me—What is a gene made of? How is it replicated? How is it turned on and off? What does it do?—would be answered within my own scientific lifetime. I had selected a topic, or series of topics, that I assumed would last out my active scientific career, and now [nineteen years later in 1966] I found myself with most of my ambitions satisfied.[43]

Growth, however, is not synonymous with change. When research findings accumulate and new observations of natural phenomenon are recorded, science grows much as a card index grows when additional works of fiction are written. Often in science, however, new findings invalidate previously held concepts, theories, and methods. As a result, change occurs. Cards are not simply added to the index; rather, they replace existing cards.[44] The core of knowledge in a field has been altered, even though the size of the field may not have changed. Indeed, what distinguishes the sciences from the nonsciences such as the arts and humanities is precisely that this core of knowledge—what one needs to know to be at the research front—does not increase in size nearly as much as it changes. Thus, although growth is the vehicle by which change occurs, it is not the vehicle that makes science science. Rather, the distinguishing characteristic of science is that growth is accompanied by change, and change makes irrelevant much of the past work. Change renders previously held truths "forever part of the past." As the neurobiologist Rita Levi-Montalcini says, herself quoting Crick: "What everyone believed yesterday, and you believe today, only kranks will believe tomorrow."[45]

From the point of view of the individual researcher, the process of change means that much of what was learned in graduate school may lose its relevance. The scientist still possesses the knowledge, but from the point of view of the scientific community that knowledge is no longer pertinent. It is no longer effective. A new theory or technique has replaced the theory or technique learned by the scientist, or to use an expression common in science, a new theory or technique has "dealt a severe blow" to an existing one.[46] One consequence of this is that the research of the scientist who does not keep up may be viewed as irrelevant or of little interest, since it relies on a knowledge base considered, if not simply out-and-out wrong, at least less effective than the current stock of knowledge is in answering the important questions of the day.

One indication of change that characterizes science is the fact that an overwhelming majority of the references cited in a typical published article were themselves published within the past five years. This is not the case for publications in the humanities. In fact, the percentage of references to materials published within the past five years is often used as an index of the "immediacy" of science. Price reports, for example, that more than 70 percent of the references in *Physical Review Letters* were dated within the last five years,[47] whereas in journals such as *German Review, American Review, American Literature, Studies in English Literature,* and *Isis*, the percentage of references dated within the last five years was less than 10.[48]

There are numerous analogies in the everyday world to scientific obsolescence. For example, the introduction of calculators in the early 1970s made desktop adding machines obsolete, and word processors have dealt a severe blow to typewriters. Or consider the case of the compact disc: As recently as 1983, over 95 percent of all recorded music sold in the United States was on tape or vinyl. Today, many stores no longer even stock vinyl, having replaced the product with the CD, which for a variety of reasons is considered vastly superior.[49] The same thing happens in science. New research techniques and ideas are introduced that make earlier ideas and techniques obsolete. Scientists who do not keep abreast of the change find themselves obsolete and the research they do relegated at best to the backwaters. They become, in the words of some scientists, "ditchdiggers."

The Importance of Fads in Science

The immediacy of science and the compactness of the core mean that scientists must keep up if they desire to be at the research front. What constitutes the research front, however, is sometimes fuzzy and malleable, being in the hands of the elite of the scientific community—the journal editors, peer review groups, and funding agencies. Often these fronts are dictated by where and how the scientific community believes the next major breakthrough is likely to occur. By the early 1950s, for example, there was a great deal of interest in discovering the structure of DNA. In his autobiography, Crick is quick to point out that one reason Watson and he received so much recognition was precisely because of what they discovered, not because of the discovery per se.[50] The structure, in his terms, made them. It was not only that they discovered the "key to life"; the scientific community was ripe, so to speak, for a solution to this puzzle.

What constitutes the research front is also dictated by fads, and some fields are clearly more susceptible than others. One area of science that is particularly vulnerable to fashion or fad is theoretical particle physics, or what is also referred to as high-energy physics. As one particle physicist told us, "Since the 1950s and possibly before that, particle theory has been heavily overlaid with transitory but highly publicized changes. Examples are Regge poles, scattering amplitudes as functions of several complex variables, dual modes, hadronic string models." Another well-placed observer in the field of physics described particle physics as "very [much] characterized by fashion." The observer continued by noting that particle physicists who do not stay in the forefront of fashion could become obsolete.

Where fashion is important, it is not predetermined that a "bad" vintage will always be bad. Styles come and go but sometimes, as in the case of miniskirts, they come again. In particle physics, for example, field theory was severely downgraded in the late 1950s and the early 1960s, after the introduction of S-matrix (scattering matrix) theory, and then enjoyed a resurgence as a result of the 1967 work of Steven Weinberg and Abdus Salam on electroweak unification. At approximately the same time, S-matrix theory began to decline in popularity, only to come back into fashion in the late 1980s when superstring theory joined some S-matrix methods with field theory.[51]

Brain research is another area in which fashion is important, particularly with regard to a theory of the human mind:

In the past few decades there have been fashions for anatomy (left brain–right brain, or the localisation of language in Broca's and Wernicke's area), for neurochemistry (the discovery in the 1970s of neuropeptides that seemed to affect sleep, pain and mood), and more recently for computer simulations of how parts of the brain work.[52]

There is a prediction that in the 1990s a different fashion will once again sweep the field, that of "studying genetic mutations, introduced into mice, which seem to knock out specific brain functions."[53]

The field of genetics is also subject to fashion in determining what work is considered at the forefront of the field. Barbara McClintock's discoveries that "some genes control other genes, switching them on and off depending on circumstances, and that some genes can change their position on the chromosome, 'jumping' from place to place" date from her experiments done in the 1940s using maize plants.[54] Yet her experiments were virtually ignored for more than thirty years. The new breed of biologists of her day, formerly physicists and chemists, worked with smaller components of the cell. For them, "simple bacteria, phage and experiments that could be done in the test tube, with cultures that reproduced every few minutes, not once a year, were not only fashionable, but proving invaluable in unravelling the secrets of life."[55] Working with complex organisms such as maize plants was simply unfashionable. In fact, it was not until the late 1970s that the value of McClintock's work was first appreciated and then honored ultimately with the awarding of the Nobel Prize in 1983.

Codification

In some fields of science, there is strong consensus concerning the questions to be asked as well as the approaches to be used. Sociologists label such fields highly "codified." Codification here "refers to the consolidation of empirical knowledge into succinct and interdependent theoretical formulations."[56] In these terms, physics and chemistry are seen as highly codified, whereas botany and zoology are not. In highly codified fields, the research front is clearly delineated. It may change rapidly because of changes in fashion, but it is delineated.

One example of a field considered to be highly codified is particle physics. Elementary particle physics focuses on the smallest bits of matter that are known to exist. Research, particularly at the theoretical level, looks for the laws governing the four forces—nuclear (strong), electromagnetic, weak, and gravitational—that describe elementary particles and their mutual interactions. The final aim of such research is to unify these interactions by finding some common origin. Abstract theorists working on unification are often depicted as involved in a pseudoreligious quest, handed them by Einstein, or, as is commonly stated in the particle literature, the search for the "holy grail." In the words of Nobel laureate Steven Weinberg:

> One of man's enduring hopes has been to find a few simple general laws that would explain why nature, with all its seeming complexity and variety, is the way it is. At the present moment the closest we can come to a unified view of nature is a description in terms of elementary particles and their mutual interactions.[57]

Although written in 1974—aeons ago in the world of high-energy physics—the search for a completely unified theory of the universe is still, in the words of Stephen Hawking, "the prime goal of physics today."[58]

In some other fields of science, there is less consensus. At the research front, widely divergent opinions are held by individuals and groups belonging to different schools of thought. Consequently, different research traditions coexist. Hull describes the "clash of doctrines" between two research groups in the field of numerical taxonomy: the pheneticists and the cladists.[59] The earth sciences are another example of a less codified discipline that permits a conceptual pluralism that is relatively rare in physics or chemistry. For the most part, the earth sciences are complex observational sciences, and as a result, "a wide range of methods (and their implied theories) may all be useful in characterizing the phenomena in question."[60]

One might think that keeping up is relatively easier in a highly codified field because the scientist does not have to guess where change is occurring. Rather, the area of change is well defined. But a corollary of the "hardness of science"[61] is its capability of growing (and, presumably changing) more rapidly than less codified or "softer" sciences do. The upside of this is that in highly codified fields, scientists need not wade through years of research to master the core; the downside is that the core can change very quickly. As a result, the threat of obsolescence is greater in such fields than it is in less codified fields. In highly codified fields, scientists are more likely to find "all their eggs in one basket" than they are in less codified fields. That is, their scientific portfolio is less diversified.

There is a second reason that the threat of obsolescence is likely to be greater in a more codified field. Not all scientists can be at the research front, although most scientists like to think that their work is at the cutting edge. In less codified fields, because there is less consensus about what constitutes the research front, there is room for a greater pluralism. A corollary of this is that in less codified fields there is less consensus about what constitutes backwater or peripheral work, because what might be considered of secondary importance from one group's perspective need not be so from another group's perspective. The risk of obsolescence is, therefore, lower in less codified fields. If you are not focusing on what some of your peers view as the most pressing problem of the day, using the techniques and concepts deemed appropriate, in less codified fields your work may nevertheless be appreciated and recognized by members of a competing research program who share your perspective.

How Scientists Cope with Change

Scientists use a variety of strategies to cope with change and the accompanying threat of obsolescence. Some of these strategies are more successful in certain fields than in others, and some are more manageable at certain types of institutions than at others. In some instances the strategy is conscious; in other instances the strategy has become so much a part of the research process that the scientist may not even think of it as a defensive strategy.

Keeping Up

An obvious strategy is to "keep up," to stay abreast of recent developments by reading preprints, attending conferences, talking with colleagues. This strategy is easier for scientists located at major research institutions where keeping up is part of doing science. Major researchers in the scientist's field pass through the doors of such institutions on a regular basis; colleagues are on national panels; and preprints flow in from around the world as competitors stake their claims for priority. More generally, scientists at research institutions are likely to be a part of the "invisible college" in their field,[62] participating in informal "continuing education" programs reserved for the elite.[63] Even for these scientists, however, keeping up is not easy. As Henry Rosovsky, former dean of Harvard's Faculty of Arts and Sciences, says, "In universities that emphasize research . . . keeping up to date can be extremely demanding and time-consuming."[64]

For scientists in peripheral institutions, keeping up is even more difficult. In order to interact with those doing work on the forefront, it is often necessary to leave the institution. To learn new techniques, one may have to visit the laboratories of other scientists working at research centers. Preprints arrive less frequently and less promptly. There is less need to send results via electronic mail to scientists at peripheral institutions, since the threat of anticipation from them is low. The matter is made worse by the fact that the process of keeping up requires resources that scientists at peripheral institutions often lack.

Keeping up requires both resources and time. Consequently, scientists who keep up risk having less time for research. A problem for scientists is to balance the two, not having their research time cannibalized by the very act of keeping up. This may be a particular problem for scientists whose work involves computers. As anyone who has ever worked with computers knows, they can be seductive. Several scientists that we questioned spoke of colleagues who devote almost all of their time to keeping themselves computer literate; that is, they become computer "junkies." Such scientists may ward off obsolescence, but they have virtually no time left for research. Their obsession with computers has, to quote one scientist, "destroyed" their productivity.[65]

The ability of scientists to keep abreast of change depends on their cognitive resources and their location within the scientific community. Although the particle physicist we quoted earlier on fashion declared that "the bright, quick theorists simply shift with the fashion," not all scientists are willing or able to "simply shift" or keep pace with fields advancing very rapidly. Depending on the quality of their vintage, scientists incur different costs in "upgrading" their stocks of human capital.

Because of what Jacob Mincer calls the "secular progress of knowledge,"[66] there is a general presumption in science that the latest educated are the best educated, at least in terms of theory and technique. Thus, in general, one would expect later vintages to incur lower costs of staying current than earlier vintages. Not only do they have to learn less to stay current. New vintages may also find it easier to keep up because newer vintages are better trained to tackle new areas. In the words of Hans Bethe, a Nobel laureate in physics:

They have been exposed to the latest points of view presented by various specialists, and now have a chance to synthesize all this information pursuing the newer lines of research. The older scientist [coming from an earlier vintage] may have the advantage of long experience, but his continuing process of self-education is more likely to be piecemeal and limited to the confines of his immediate working interests. It is not uncommon for him to realize that his students may know more about many areas of his wider field than he does.[67]

This type of reasoning suggests that independent of location, scientists from more recent vintages are more likely to learn the new methods and techniques than are scientists from earlier vintages. Not only are the costs likely to be lower, but because more years are left in their career, the benefits to learning the new theories and techniques are likely to be greater.

Consider an analogy with physical capital, the introduction of personal computers (PCs) in the workplace in the early 1980s. PCs clearly revolutionized data processing and information retrieval. As people became acquainted with their capabilities, PCs rendered whole categories of equipment obsolete, such as desktop adding machines and, to a large extent, even typewriters. Not only was the old equipment too slow, but the finished product was often not as good or accurate as that obtained with the new PCs.

But the microcomputer revolution has evolutionary aspects as well; there have been many small changes since PCs were initially introduced, and even these changes impose costs on keeping up. Software and hardware have continually evolved, and the 8088 chip of the early PCs is either too slow—PCs designed for the 1990s process twenty to thirty times faster than the original vintage does—or does not have the capability to process today's programs. Depending on the machine's vintage, upgrading may be possible. But the costs of doing so generally rise with the lapse of time. That is, the costs of upgrading are usually higher for earlier models, and at some point, it simply does not pay to invest in upgrading. Indeed, it may not even be possible to attain current standards via the upgrading route.

A similar story may be told for human capital. In certain cases, learning new techniques and theories, that is, upgrading, may simply be too expensive, and scientists may choose not to keep up with work on the frontier. There are a variety of strategies that such scientists can follow. One is to do backwater or sidewater research. Another is to switch fields. Scientists who do not upgrade may also move into administrative positions or choose to focus exclusively on teaching. Some may become historians of science, and others may write textbooks.

Backwaters and Sidewaters
Scientists who continue to work in the same area after major advances have taken place are often described as engaging in backwater research. The success of this strategy[68] hinges in part on the field in which the scientist is working. In some areas, such as particle physics, backwater research approaches being an oxymoron. The cutting edge is so closely integrated with the "received" core of knowledge in the field that little else is acceptable. In other areas, however, such as solid-state/condensed-matter physics, backwater research is possible. Although they are likely to lose substantial prestige in the community, scientists who do not learn the new theory or technique can still engage in research that is of some interest to the community. In still other areas, there is room for a more pluralistic approach, and so working in the sidewaters is more acceptable. For example, in the earth sciences, there is little reason to think that the conceptual revolution of plate tectonics has compromised scientists' ability to remain productive. Research can still proceed in the usual manner in this largely observational field. This is true partly because the earth sciences, unlike physics (and especially particle physics), are not as "codified"; hence a wider range of methods and approaches are acceptable to the scientific community.[69]

It is important to emphasize that there are some advantages, both to the scientist and to society, to engaging in backwater and sidewater research. Scientists doing such research may be less encumbered by the standards of the new (received) conventional wisdom. Consequently they may have different insights and may be more willing to challenge the new truths. As a result, major breakthroughs in science can result from such research.[70] We have already noted the example of Barbara McClintock's path-breaking work in genetics which for years was considered to be, if not in the backwaters, at least in the sidewaters of genetics. Most people would agree that the work on chaos came out of the backwaters of mathematics and physics.

More generally, it can be argued that an obsession with path-breaking research, often at the expense of other types of research, presents problems for the United States. Applied research is the mechanism that translates basic research into economic growth. Perhaps the United States would do well to encourage more scientists to engage in work in the backwaters and sidewaters, looking for applications that come from the core of scientific knowledge. Japan has certainly done well by fishing in these waters.

A Shift of Focus

Another way that scientists cope with change is to shift the focus of their work. For some, this means switching fields. A common pattern in physics, for example, is for scientists to move from particle, considered the most "basic" of all areas of inquiry, into a more applied area such as solid state/condensed matter. Scientists who make such a move generally lose status, however, since particle physicists see themselves as being at the pinnacle of a well-established pecking order of prestige. Nevertheless, they are likely to continue to be productive for a long time. Other scientists threatened by change, particularly those with the gift to write, can become chroniclers of science, drawing the "big picture" for the educated public. Still others may choose to devote their energies to writing textbooks. Indeed, a physicist suggested to us that the best way to determine who is obsolete is to ask who is writing textbooks!

Another career move available to scientists is to go into administration: chairing a department, running a lab, becoming president of an institution. Some scientists move on to the role of science-statesman, having a significant impact on the allocation of resources in the scientific community. Scientists in such positions usually claim that they no longer do research because they have no time, but others contend that they do no research because they no longer have "any science left in them."[71]

Another way that scientists can move away from research in the face of obsolescence is to concentrate exclusively on teaching. Scientists choosing such a course, however, may find that they no longer know as much as some of their students do. They may also feel a sense of alienation from the field and a substantial loss of prestige. But their jobs in the classroom are secure. They have tenure. This option, however, is generally open only to scientists who already have jobs in the academic sector, for, as we noted earlier, scientists generally do not move from careers in business and industry or government into the academic sector.

ARE TODAY'S SCIENTISTS AT INCREASED RISK?

One important feature of science in the United States is that in recent years the scientific enterprise has become increasingly weighted toward scientists educated in earlier times. This can be seen by looking at the distribution of Ph.D.'s in

1973 (the earliest year for which data are available) and 1987 (the latest year for which comparable figures are available.) For convenience, scientists are grouped according to when they received their Ph.D.'s. Table 6-2 shows that in recent years, because of the slowed growth in science, scientists educated at a later date have made up a smaller and smaller percentage of the population of practicing scientists in the earth, life, and physical sciences. In 1973, for example, in these fields over half of all Ph.D.'s had been awarded within the past ten years. By 1987, after a period of slowed growth, this figure had fallen considerably. Especially affected were the physical sciences, in which the percentage plummeted from 50.7 in 1973 to 31.1 in 1987.

Although it is not always true that scientists educated more recently are better prepared than are those educated earlier or that those educated more recently witness less change in their fields, this is in fact generally the case. Thus, the percentages of Table 6-2 stand as a chilling reminder of another consequence of the slowed growth that began in science in the early 1970s in the United States.

The fact that problems accompany change does not mean that change in science is not valued. Change is a double-edged sword. Without change, science would not be science; we would have no pasteurization of milk, no ability to circumnavigate the world in a matter of hours, no knowledge of the universe. Change, the belief in change, is what attracts individuals to science. Scientists are not just explorers, charting unknown waters. Rather, the goal of scientists is to bring back ideas and techniques from their explorations—whether they be into the interior of the atom, the depths of the earth, or the furthermost galaxies—that change the way we think. But innovations and discoveries in science take their toll on the productivity of those scientists not intimately involved in the process of change. Consequently, change is both the goal and the nemesis of scientists.

For several reasons, it could be argued that scientists in the United States

Table 6-2. Percentage Distribution of Scientists by Length of Time Since Ph.D. Received, 1973 and 1987

Year of Ph.D.[a]	Earth sciences		Life sciences		Physical sciences	
	1973	1987	1973	1987	1973	1987
Last 5 years	35.9	20.4	33.3	19.0	29.4	14.6
Last 6 to 10 years	24.2	22.8	21.1	22.4	21.3	16.5
Last 11 to 15 years	14.4	23.4	14.4	22.2	15.0	20.2
Last 16 to 20 years	11.5	16.4	14.4	16.7	13.9	21.2
Last 21 to 25 years	6.8	10.5	7.9	8.6	9.8	11.9
Last 26 or more years	7.3	6.4	8.8	11.1	10.7	15.6

[a] Given the timing of the SDR, the distributions begin counting (backwards) from the calendar year two years earlier than the survey date; for example, the last five years for the 1973 SDR includes 1967 to 1971.

Source: Survey of Doctorate Recipients (SDR) tabulations.

today are more adversely affected by change than scientists have been at earlier periods of time. First, as Table 6-2 demonstrates, scientists in the United States today come, on average, from earlier vintages than they did in previous periods. Although later vintages, in terms of their knowledge base, need not always be better, this is likely to be the case. Furthermore, as we have contended, earlier vintages with a knowledge disadvantage may have more difficulty keeping up than later vintages do. Second, relatively speaking, more scientists are located in peripheral institutions today than in the past. This is especially true of scientists educated in more recent years in the physical sciences. Assuming that change persists in science, these newer vintages may be hard pressed to keep up in the years to come, working in environments in which keeping up is exceptionally difficult. Finally, keeping up requires resources to which the average scientist has less access today than in years past. Not only has the rate of growth of expenditures for research in the private and public sector slowed, but the resources that are spent, particularly at the federal level, also have centered more and more on expensive projects like the Hubble telescope and the human genome-mapping effort.[72] Such projects, even if 100 percent successful, will mean that average scientists have less access to resources than they did a decade or two earlier and so will create a "mad scramble for a few dollars."[73]

SUMMARY

In this chapter we examined the importance of RPRT for a scientist. We argued that job market conditions that greet newly trained scientists as well as the intellectual climate existing at the time that they received their training can have a profound effect on their career path. Scientists educated when jobs are abundant in the research sector have an edge that others do not have. They are likely to settle at an institution that nurtures research. Likewise, scientists trained at a time when change is occurring in their field have a research edge that others do not have. Particularly favored are those who train at the place where the change begins.

RPRT means that success in science depends in part on conditions outside the individual's control. It also means that certain generations of scientists may be more productive than other generations, since much of RPRT is keyed to when a scientist was trained. Scientists educated at approximately the same time share a knowledge base more similar than the knowledge base shared with others. They wear an imprint of the science of the day, embodying the know-how and research style transmitted to them while they were in graduate school. They are, loosely speaking, of the same vintage. Scientists trained at the same time also share common experiences in regard to finding jobs in the research sector. Some Ph.D. classes graduate when vacancies are plentiful and jobs are there for the picking. Others complete their training at a time when it is almost impossible to locate at a top research institution. In the next chapter we explore yet another reason for generational differences.

NOTES

1. His own experiments with the multiplication of bacterial viruses called *bacteriophages*, or *phages* for short, eventually earned him the Nobel Prize in 1969.
2. The "phage group" was an "invisible college" of researchers from many different areas investigating the mystery of the genetic material. They met each summer at Cold Spring Harbor, Long Island.
3. By the age of 31 (by 1969), Snyder was already a full professor of pharmacology and psychiatry at Johns Hopkins and the winner of several prestigious awards. The discussion in this paragraph draws heavily on the work of Kanigel 1986.
4. Kanigel 1986, p. 168.
5. The term *molecular psychology* was coined by Jon Franklin; see Franklin 1987.
6. See, for example, work by Blackburn, Behymer, and Hall 1978; Blau 1973; Long 1978; Long and McGinnis 1981; Pelz and Andrews 1976.
7. The statistics refer to those scientists trained in a particular discipline who are still employed in that discipline. The specialties of astronomy and astrophysics were not included in the definition of the physics discipline (see Stephan and Levin 1987).
8. Salamon, as quoted in G. Cole 1979, p. 361.
9. Lodahl and Gordon 1972.
10. Wrong place, wrong time does not apply only to scientists. A recent story in *The Economist*, March 2, 1991, recapping the Persian Gulf War, relates how a reporter who was a soldier in Vietnam and a journalist in Saudi Arabia came to the realization that he was in the wrong place at both times.
11. Porter 1979b, p. 133.
12. Alpher et al. 1979, p. 27.
13. Long, Allison, and McGinnis 1979, p. 816. Traweek 1988, p. 92, discusses how particle physicists flow from the "core" to jobs in the "periphery." She argues that this pattern is particularly strong at the postdoctorate level: "I know of no cases in which a particle physicist moved from a postdoc at a peripheral place to an established position at a core institution."
14. National Science Board 1987, Appendix Table 2-11, p. 204.
15. The job market for scientists is often described by economists as fitting a cobweb model. In such a model, supply at one point in time is largely determined by job market conditions existing several periods earlier. Freeman 1975, for example, discusses the operation of such a model.
16. These rates of increase were calculated from special tabulations from the Doctorate Records File made available to us by the National Research Council. The "sciences" include the physical sciences; the earth, environmental, and marine sciences (what we call the earth sciences); the mathematical sciences; and the life sciences. Engineering and the social sciences are excluded.
17. This was calculated for the period 1920 to 1974 using data from National Research Council 1978, Table 2A, p. 12.
18. Federal support was provided by such agencies as the National Science Foundation, the National Institutes of Health, and the National Aeronautics and Space Administration; see Bowen and Sosa 1989, p. 177. Also important as a source of funds for graduate support was the training grants program created by the National Defense Education Act of 1958. The number of graduate students (in all fields) on federal fellowships and traineeships rose from under ten thousand to over fifty thousand between 1959 and 1967 (see U.S. Congress 1989, pp. 126-7).
19. Ahlburg, Crimmins, and Easterlin (1981) offer evidence on the impact of the draft on college enrollment rates of men.
20. U.S. Congress 1989, p. 127.
21. National Research Council 1978, p. 95.
22. Bowen and Schuster 1986, p. 91.
23. Koch 1971, p. 24.
24. For an example of a research agenda inspired by the war on drugs, see Snyder 1989.
25. This information was added to the Doctorate Records File beginning in 1958. The percentages reported were specially tabulated for us by the National Research Council. These statistics must be viewed with some caution, however, primarily because Ph.D. recipients provide the infor-

mation anywhere from six months prior to graduation to six months after graduation. In addition, during the last thirty years, the definitions used to classify the Ph.D.'s have changed somewhat.

26. The data are from the Survey of Earned Doctorates and are available in the 1988 summary report. The percentages are calculated for those who indicated on the survey that they have a definite commitment. Note that the increase in the percentage of new Ph.D.'s taking postdoctoral appointments is indicative not only of the state of the job market but also the increasing number of foreign graduates, since temporary residents find it easier to obtain postdoctoral positions than "permanent" positions.

27. There are two major databases for Ph.D. scientists: The Doctorate Records File contains information from all Ph.D. recipients in the United States at the time they received their Ph.D.'s. (Since 1957, all Ph.D. recipients have been required to complete the Survey of Earned Doctorates.) Because many Ph.D.'s take postdoctoral appointments, employment statistics for this group in terms of the categories of Table 6-1 are not very useful. The other database is the biennial Survey of Doctorate Recipients. Unfortunately, because it was not begun until 1973, it contains little information on the success of the 1960s generation in locating their first job in the research sector.

28. Doctoral recipients in the social sciences and humanities also were affected by the market's boom-and-bust cycle and may have fared even worse than did their counterparts obtaining doctoral degrees in the sciences. Indeed, one speaks of an entire "lost generation" who received doctoral degrees, but not tenure-track academic appointments, in the 1970s and early 1980s (see Scott Heller, "The Expected Turnaround in the Faculty Job Market May Come Too Late for 'Lost Generation' of Scholars," *Chronicle of Higher Education*, May 23, 1990).

29. See Rosovsky 1990, p. 155. The rigidity between sectors can be seen by examining the proportion of scientists in the nonacademic sector who move into the academic sector. Ehrenberg and his colleagues (1991) report, for example, that between 1985 and 1987 less than 3 percent of all physical scientists who had been working in the nonacademic sector moved into the academic sector. In environmental sciences, the figure was closer to 5 percent. Mobility from industry and government into academics is facilitated, however, by a tight labor market. For example, Ehrenberg found that in 1987, approximately 20 percent of all new academic positions in engineering schools were filled by persons coming from industry or government.

30. The data prepared by Harmon (National Research Council 1965, p. 10) shows that the choice of a nonacademic career is often final. Work by Caplow and McGee (1958) also suggests that after employment in industry, it is almost impossible to return to academia.

31. Research universities include institutions of higher education classified as either Research I, Research II, Doctoral I, or Doctoral II by the Carnegie Foundation (see Carnegie Foundation for the Advancement of Teaching 1987).

32. If anything, the percentage of the earlier cohorts still employed in the top research sector as of 1987 should be a lower-bound estimate of the percentage initially acquiring positions in this sector, since job moves within the hierarchy of the scientific community are more likely to be downward than upward.

33. The table stops with 1982 because information on the most recent graduates from the 1987 survey is not comparable, since many recent graduates hold postdoctoral positions.

34. Menard 1971, p. 24.

35. Ibid.

36. Quantum field theory is an attempt to explain electric and magnetic forces as a quantum theory. In solid-state physics the techniques of quantum field theory were found to be useful because electrons, when excited, depart from their normal state and behave much like particles. The behavior of these so-called quasi particles can be analyzed mathematically as a system of infinitely many interacting particles: Hence the name many body theory.

37. Cohen 1985, p. 10.

38. Snyder 1989, p. 189. Snyder studied with Julius Axelrod, a mentor to several Nobel Prize winners. A similar theme is also expressed by Kanigel 1986.

39. Traweek 1988, p. 17.

40. Ibid., pp. 75–76.

41. Feynman 1985, p. 226.

42. See Price 1986, p. 1, for the percentage of scientists still alive; p. 5, for the literature growth

rate; and p. 11, for population. Note that one consequence of exponential growth in science is that it is not just true that 80 to 90 percent of all scientists who have ever lived are alive today, but as Price (p. 12) so eloquently states, "It follows that this result, true now, must also have been true at all times in the past, back to the eighteenth century and perhaps even as far back as the late seventeenth."

43. Crick 1988, p. 143.
44. Sometimes, with the emergence of a new field, a whole new index box is required.
45. Levi-Montalcini 1988, p. 123.
46. See ibid. for an example of this.
47. Price 1970, p. 21.
48. Ibid., p. 15.
49. See Gerry Yandel, "Record Stores Face the Music: Albums Doomed to Extinction," *Atlanta Journal Constitution*, October 25, 1988. In 222 Camelot Stores throughout the United States, LPs represented 41 percent of sales in 1984, CDs 4 percent. In 1989 the comparable figures were 40 percent for CDs, 3 percent for LPs.
50. Crick 1988, p. 76, says, "Rather than believe that Watson and Crick made the DNA structure, I would rather stress that the structure made Watson and Crick." Crick stresses (p. 74) that "the major credit I think Jim [Watson] and I deserve, considering how early we were in our research careers, is for selecting the right problem and sticking to it. It's true that by blundering about we stumbled on gold, but the fact remains that we were looking for gold."
51. See Pais 1986 for a discussion of developments in particle physics.
52. See "A Survey of Science," in a special supplement to *The Economist*, February 16, 1991, p. 9.
53. See ibid.
54. Gribbin 1987, p. 78.
55. Ibid., p. 302.
56. Zuckerman and Merton 1973, p. 507.
57. Weinberg 1974, p. 50.
58. Hawking 1990, p. 69.
59. Hull 1988.
60. Law 1980, p. 17.
61. This term, originally coined by Storer, refers not to the difficulty of science but, rather, to the fact that scientific knowledge, as opposed to humanistic inquiries, is "brittle, unyielding, impersonal, and hard" (Price 1970, p. 4).
62. The concept of invisible colleges was discussed extensively by Price 1986, and also Crane 1972.
63. For example, work in the fledgling field of microbiology was greatly facilitated by participating in the "phage group" that met each summer at Cold Spring Harbor, Long Island (see Gribbin 1987, p. 221).
64. Rosovsky 1990, p. 162.
65. Another physicist told us, "I am not convinced that those physicists who never learn the trade [computers] will ever suffer from it in their productivity. Indeed, the correlation may go the other way: those who invest large amounts of time learning computers may actually suffer in terms of physics publications."
66. Mincer 1974, p. 21.
67. As quoted in Reif and Strauss 1965, p. 302.
68. Backwater research may not always be a survival strategy adopted in the face of change. Some scientists do backwater research out of their belief that their field has taken a false turn and will eventually return to their point of view.
69. Law 1980.
70. This relates to the discussion of marginality and innovation presented in Chapter 3.
71. Traweek 1988, p. 102.
72. William J. Broad, "Vast Sums for New Discoveries Pose a Threat to Basic Science," *New York Times*, May 27, 1990.
73. Ibid. The "crisis" in funding is discussed in more depth in Chapter 9.

7
Quality in Science

Everyone knows that model matters when it comes to choosing an automobile. Some models are lemons, some workhorses, some just average. Some have a sense of style that transcends time; some have characteristics that mark them forever as mistakes, better to be forgotten. Consider the Ford Mustang. To date, there have been three distinct models: the original "pony," built from 1964 to 1973; the Mustang II, built between 1973 and 1979; and the current Mustang. Given the strong consensus that the Mustang II was an underwhelming car, whereas the pony was a car that stole your heart, a classified advertisement for a 1970s Mustang will attract much more attention if the car was built before 1973. The reason is model, not age. Or consider wine. By all accounts, 1979 was an exceptional year for white burgundies, and 1984 was one of several mediocre years in the 1980s. Today 1979 white burgundies command a higher price than 1984 burgundies do. The reason is vintage, not age.

RPRT means that model matters in science, that, other things being equal, certain models of scientists are more productive, on average, than other models are. In this chapter we explore another reason that certain models of scientists may be more productive than other models are. Here, rather than concentrating on RPRT, we look at the idea that the "quality" of the scientific community is related to model and consequently, other things being equal, that certain models may have more science in them than other models do. Before turning to quality considerations, we define the concept of cohort.

WHAT IS A COHORT?

Demographers use the term *cohort* to stand for those persons "who pass some crucial stage at approximately the same time, like marriage, first employment, and especially birth."[1] These individuals, coming of age or into a particular status or condition at the same time, share similar experiences or receive similar exposure to the unique events that characterize their lives. These common experiences may mark each one for life, and as a result, members of different cohorts may exhibit differences in behavior, values, and intellectual abilities.

A popular example is the sixties generation, a generation whose coming of age was shaped by the tumultuous events of the late 1960s—the Vietnam War and the antiwar movement, the civil rights movement, drugs and spiritual quests, new forms of love and work, radical politics, and the rise of feminism. Even though two "waves" of this generation can be identified (the first being those born between 1944 and 1949, the second between 1950 and 1957), Annie Gottlieb concludes in her study of this generation that "the two 'waves' of the Sixties generation are much more like each other than they are like the generations before and after them, who were too old or too young to be intimately shaped by the Sixties."[2]

In addition to being exposed to different values and world events, cohorts often differ in terms of size. The baby-boom generation of the 1950s and early 1960s, for example, was much larger, in both absolute numbers and relative size, than the baby-bust generation of the 1970s. Social scientists have studied the effects of size of birth cohort on a host of factors, including various aspects of child development such as mental and physical health, educational achievement, intelligence, and personality.[3] Social scientists have also studied how certain aspects of adult behavior such as fertility, labor force participation, and earnings are related to the size of the birth cohort. A consistent finding in this body of work is that "other things being constant, the economic and social fortunes of a cohort tend to vary inversely with its relative size."[4] Small is better—at least from the perspective of individual members of the cohort. The reason is not hard to fathom, especially in the adult years, when cohorts that are relatively small have an advantage at the time of entry into the labor force, enjoying relatively favorable economic circumstances in terms of wages, employment, and the prospects of upward mobility. Moreover, as Richard Easterlin and others who have studied the issue have demonstrated, these favorable conditions tend to foster somewhat higher levels of personal satisfaction and higher rates of fertility, as well as earlier ages of marriage and retirement for the advantaged cohort.

From our perspective, it is not the birth year but the year of the Ph.D. that serves to index the distinctive experiences of different cohorts entering science. Two ways in which scientists' careers can be shaped by events occurring when they were in graduate school were discussed in the last chapter. Here we explore why different cohorts of scientists may also differ in certain qualities related to scientific productivity—attributes such as talent, motivation, and values. Combined with the forces of RPRT, these quality considerations provide a third reason that productivity can be related to cohort.

QUALITY AND COHORT

A common theme in science today is that "young people don't have what it takes."[5] There are several reasons, all cohort related, to believe that this is not simply nostalgia but that the quality composition of the scientific community actually has changed over time. Here we examine both the reasons and the supporting evidence for quality differences among cohorts.

The "selectivity" hypothesis, which can be traced to Derek de Solla Price, claims that because of the rapid growth experienced in science, the United States is "scraping the bottom of the barrel" of the talent pool, that science has become less selective over time.[6] A related theme is that in order to attract more and more people into science, it has become necessary to secularize science, placing more emphasis on financial inducements. Consequently, it is argued that scientists today do not have the same level of motivation as they had when the primary attraction of science was its power to answer questions.

A second argument, the "brain drain" hypothesis, focuses on the possibility that cohort differences in the quality of the scientific talent pool exist because the relative attractiveness of careers in science changes over time. First, the opportunity structure shifts. As Michael Sovern, president of Columbia University, writing in 1989, put it:

A generation ago many students graduating at the heads of their class were convinced that an academic career represented their finest opportunity. . . . [Today,] the choices of potential Ph.D. candidates are clear: Go to work at once, go to law or business [or medical school] and earn a high salary in a few years, or spend a longer time earning a Ph.D. to earn less as a professor. It is remarkable that over the years so many people have made that last choice, but fewer of the best have been making it lately.[7]

Second, societal values and interests in science change over time. "Twenty-five years ago the followers of Darwin and Einstein . . . were admiringly called whiz kids. Now we have geeks, nerds, and dweebs."[8]

A third argument, the "sociocultural" hypothesis, is that there may be generational differences in average ability, creativity, and motivation attributable to sociocultural factors. One such factor is crowding. The Easterlin hypothesis mentioned earlier contends that compared with other cohorts, the baby boomers are disadvantaged in all aspects of their lives because they have encountered crowding in the family, school, and workplace. A second factor, championed by Dean Keith Simonton, is that differences in the incidence of certain sociocultural and psychological events such as wars lead to generational differences in individual scientific creativity.[9] A similar theme is expressed by Joel Mokyr in his recent study of why some societies are technologically more creative than others are.[10]

The Selectivity Hypothesis

In *Little Science, Big Science*, written in 1963, Derek de Solla Price[11] heralded the transition in science from a state of roughly exponential growth to a mature state, characterized by a continual decline in the rate of growth. Using the analogy of a bean plant, Price pointed out that exponential growth cannot continue forever unchecked, that eventually limits to growth are reached and a mature phase ensues in which the bean plant, or in this case science, remains at a fairly constant size.

The exponential growth that Price documents is (depending on how one measures science) between 5 and 7 percent, meaning that science has been doubling every ten to fifteen years. Although the world has been able to sustain this growth rate for at least three centuries, Price (among others) asserted that such a rate cannot persist forever, and that the saturation point will be reached in the not-so-distant future. Indeed, as Price so aptly stated, if the growth rate were to remain the same, in less than a century there would be "two scientists for every man, woman, child, and dog in the population and we should spend on them twice as much money as we had."[12]

Price's concern with growth can be easily understood by computing the number of U.S. Ph.D.'s in science and engineering per one million persons in the United States 25 years and older. If this number stays constant over time, science will grow at the same rate as the population does, but if it increases, growth in science will outpace that of the population. In 1920 this figure was a mere 82, by 1930 it had doubled to 169, and by 1940 the density had risen further to 320. Although relative growth was slowed during the war years, by 1960 there were 778 science and engineering Ph.D.'s per million persons 25 years and over living in the United States, and during the next decade, which witnessed the extraordinary period of growth that we noted earlier, the density almost doubled to 1507. Ten years later, the ratio had increased even more, and by 1987 there were approximately 2089 Ph.D. scientists and engineers per million persons 25 years and over living in the United States. Thus, over a period of sixty-seven years, the Ph.D.-to-population ratio grew by a factor of more than 25.[13] Clearly, growth in science has overwhelmed growth in the population. Although the United States is far from saturation, it has had to dig deeper into the talent pool to come up with the higher density of scientists. Furthermore, the rate of growth, as the numbers imply, has begun to slow.

Price predicted that growth in science would slow long before the saturation level was reached. It would slow not only because of the strain placed on the economy by the immense financial resources needed to keep the bean plant growing but also, and perhaps more importantly, because the number of people capable of being scientists is limited. According to Price, approximately 6 percent of the adult population possess the minimum capabilities needed for science. Furthermore, because not all of the 6 percent are equally able, growth slows long before the entire 6 percent become scientists. Indeed, Price argued that the number of average scientists increases as the square of the number of highly eminent ones does. Thus, for the eminent pool to double in size, the average pool must increase by a factor of four. Consequently, Price argued that the average quality of scientists decreases over time as successively larger and, on average, less able cohorts of scientists are added to the scientific community. "One ceases to skim off the cream; society begins to have to work against the natural distribution of talents."[14]

A simple example illustrates this point: Suppose that 6 percent of the population possess the minimum ability to be a scientist but that exceptional talent is found in only 1 percent of the population. Let the growth rate of science be twice that of the population, 100 percent compared with 50 percent, and observe in Table 7-1 what happens to the talent composition of science over time. In period A, the entire scientific labor force is drawn from the most highly talented group of individuals; the scientific community is 100 percent cream. But in period B, of the 600 scientists needed, 450 are drawn from the most highly talented group, while the remainder come from those able but not exceptional. The cream content has fallen to 75 percent. And in period C, the cream content falls further to 56.25 percent. Admittedly, the growth rates used in the example are extreme, but the point is obvious: Sustained growth is accompanied by a decline in average quality.

Some may contend that a decline in average quality is of no concern as long as the cream is still present. Such an argument, however, overlooks three salient facts. First, science as we have repeatedly pointed out, is a social enterprise. The exchange of ideas is important. And as the cream content diminishes, the quality of this interaction is likely to decline. Second, significant growth often is accompanied by increased competition for scarce research resources, and it is not at all clear, as we shall see in Chapter 9, that the best, most creative projects are always funded. Therefore, a decrease in average quality, brought about by growth, may lower the probability that top research is supported. Finally, one can argue that cream attracts cream. As science becomes less selective, the best and the brightest are less attracted to the field.

Is there any "hard" evidence that the growth in science has meant a lower average quality of persons in science? This turns out to be an especially difficult question to answer. Although there is wide consensus that some people are particularly good at doing science, there is significantly less consensus on how to go about measuring this ability, particularly since many of the dimensions involved are somewhat fuzzy. For example, one dimension relates to motivation, another to what may be called scientific intuition. Intellectual aptitude is also a factor, although, as was noted in Chapter 2, the evidence concerning a relationship between IQ and scientific productivity is sparse, perhaps because the scientists who are included in such studies are almost always above whatever threshold of intelligence is considered necessary to do science. Another reason that it is especially difficult to address the quality issue is that most of the data that exist have been collected over such a short period of time that reductions in quality, even if they occur, may be difficult to decipher. There are very few data on the ability of people who became scientists before the 1960s.

Table 7-1. Hypothetical Example of Price's "Skimming the Cream"

	Period		
	A	B	C
Number of scientists	300	600	1200
Size of population	30,000	45,000	67,500
Size of talent pool	1800	2700	4050
Number with exceptional talent	300	450	675
Percentage of scientists with exceptional talent	100%	75%	56.25%

And the period when the data began to be collected, the 1960s, coincides with a time of extraordinarily rapid growth in science. Precisely because of this rapid expansion, the talent pool may have already been diluted. Thus the benchmark for comparisons with future cohorts may be at a low level that fell very little during the next twenty years when, with the exception of the life and computer sciences, the number of people going into science grew at a snail's pace if at all.

Another problem in answering the quality question relates to the fact that in recent years there has arguably been a brain drain away from science as the best and the brightest have sought careers in other, more lucrative and attractive fields. Thus, if there is a drop in quality, it is difficult to know whether it is due to the growth effects that Price outlined or to the brain drain effects discussed next. Finally, across the board in all fields of study, standardized test scores of young people fell and then rebounded during the next fifteen years. An explanation for this, which we explore shortly, is that these trends can be explained by the cohort-size hypothesis: The lower test scores, on average, of the postwar baby-boom generation are being replaced by the higher scores of the baby-bust generation that succeeded it.

The only database on Ph.D.'s in the United States that goes back for any period of time is the Doctorate Records File (DRF) which, since its inception in 1920, has collected information on virtually every person awarded a Ph.D. in the United States.[15] Although the information collected is fairly meager, since 1931 the records have contained the undergraduate school that awarded the baccalaureate degree. Thus we can use these data to determine the percentage of scientists, from 1931 to the present, who received their undergraduate training at distinguished colleges and universities. There is, of course, ample evidence that excellent graduate students come from a wide range of colleges and universities, and it is clear that prestigious colleges and universities bestow baccalaureate degrees on individuals of less than extraordinary ability. Still, using a quality index based on the selectivity of the college or university, one would expect a strong positive correlation between such selectivity and the ability to do science.

Figure 7-1 shows the percentage of Ph.D. scientists who, during the past fifty-nine years, received their undergraduate degrees at one of sixty-five "top" undergraduate academic institutions in the United States. These top institutions are the "most" selective—with a rating of "7"—according to Alexander Astin's index of selectivity.[16] Astin's index, which is based heavily on student aptitude, includes in category seven such schools as Harvard, Yale, California Institute of Technology, Cornell, Chicago, Vanderbilt, Johns Hopkins, Carleton, Vassar, and Stanford. In order to eliminate the bias that would result from including students who received undergraduate degrees from foreign institutions, we report the percentage only for scientists who received both their undergraduate and doctoral degrees in the United States. Not surprisingly, the percentage from the most elite sector has fallen over time. For example, in 1931, over 26 percent of those receiving Ph.D.'s in that year had been trained at

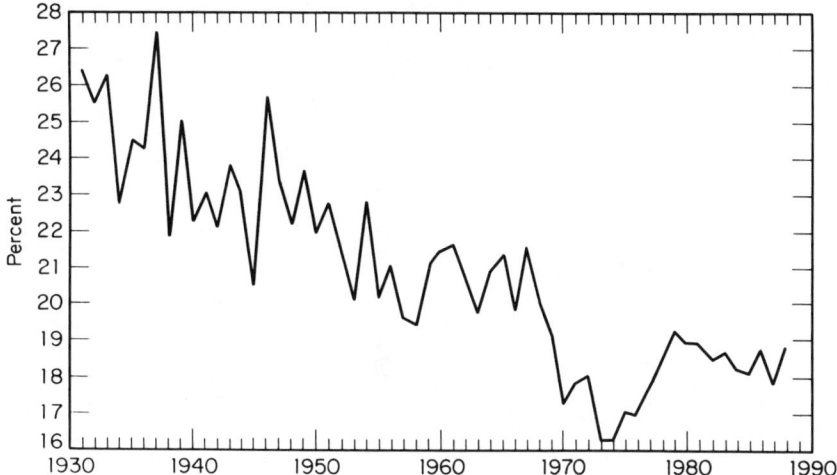

Figure 7-1. Percentage of Ph.D. recipients in the life and physical sciences with U.S. baccalaureate degrees from the "most" selective schools [Astin (1971) rating of 7]. *Source:* Doctorate Records File.

one of the top undergraduate institutions. By the late 1950s and early 1960s, this figure hovered around 20 percent, and with the rapid growth in science in the late 1960s and early 1970s, the percentage dropped further to a low of approximately 16 percent in the period 1972 to 1974.

Additional evidence concerning selectivity can be inferred by comparing the proportion of Ph.D. recipients from the most selective undergraduate institutions with the proportion from institutions of median-level selectivity, those rated a "4" by Astin. (Included in this group, for example, are the University of Arkansas, Auburn University, Calvin College, Clemson University, Coe College, the University of Missouri at Columbia, Oakland University, and the University of Oregon.) Here prudence dictates limiting the analysis to a more recent time period, since the ratings for median-level institutions are, in all likelihood, less stable over time than are the ratings for the most selective institutions.[17] Table 7-2 presents the ratio of Ph.D. recipients with baccalaureates (B.A.'s) from the most selective institutions per one hundred Ph.D. recipients with B.A.'s from institutions with median-level selectivity for the period 1959 to 1988. With rare exceptions, before 1968 more doctorates in science had trained at the most selective schools than at the median-rated schools. Since 1968, however, the situation has reversed itself. This reversal was particularly noticeable during the early 1970s when, in two instances, for each one hundred Ph.D.'s receiving their B.A.'s at median-level rated institutions, fewer than seventy-five had obtained their B.A.'s at the most selective undergraduate institutions.

The ratios are consistent with John Bishop's finding of a decline in the Graduate Record Exam scores of those applying to graduate and professional

Table 7-2. Number of Ph.D. Recipients in the Life and Physical Sciences with U.S. Baccalaureate Degrees from the "Most" Selective Schools per 100 Ph.D. Recipients from Schools with Median-Level Selectivity, 1959–88[a]

Year	Number	Year	Number	Year	Number
1959	104	1969	86	1979	90
1960	107	1970	79	1980	89
1961	104	1971	77	1981	85
1962	101	1972	82	1982	85
1963	99	1973	73	1983	85
1964	104	1974	71	1984	82
1965	111	1975	76	1985	81
1966	93	1976	73	1986	84
1967	106	1977	79	1987	79
1968	98	1978	84	1988	83

[a]Rating of 7 compared with rating of 4, see Astin 1971.
Source: Doctorate Records File.

schools during the late 1960s and 1970s.[18] He attributes the decline in average quality, at least in part, to the "substantial increase during this period in the proportion of BA recipients who entered graduate or professional schools."[19] That is, quality decreased because of growth.

The one study that directly attempted to examine Price's thesis did not support it.[20] The study's failure to measure relative cohort size, and its undemanding criterion for measuring quality, however, makes it less a refutation than an example of the pitfalls encountered in addressing the "creaming hypothesis." The study in question examined whether quality in physics is related to cohort size. Quality was measured by whether Ph.D. physicists received one or more citations to works published during the first three years of their career. Changes in cohort size were measured by changes in the number of assistant professors hired between 1963 and 1975. The authors, Stephen Cole and G. S. Meyer, found that over this period, regardless of the size of the cohort, the proportion of cited physicists remained constant at approximately 30 percent each year; hence they concluded that quality does not depend on cohort size. The very loose criterion used to measure quality and the authors' measure of relative cohort size significantly dull the study's conclusions, however. With regard to quality, Price considers that a minimal scientist is able to write one article, and the eminent can write many of high quality. Moreover, the eminent constitute no more than about 10 percent of all scientists, not the 30 percent found in this study. Cole and Meyer's definition of cohort is also deficient. Price was not talking about the percentage of scientists who enter academe, but rather about the percentage of the population who become scientists. During most of the period covered by the Cole and Meyer study, the two moved in opposite directions, with the percentage of persons receiving Ph.D.'s in physics increasing while the percentage of physicists obtaining academic appointments was decreasing.

The Secularization of Science

Price was concerned not only that the talent of the scientific community may decrease with growth but also that Big Science may produce "a breed of scientist different from that of Little Science."[21] When science was smaller, according to Price, many scientists chose science because of the emotional gratification that its puzzle-solving challenge provided. During this phase, according to Menard, science was "like a fundamentalist religion in size and dedication."[22] Growth has made science big, and to attract more people into science it has become necessary to secularize the reward structure. Now, according to Price, it is not only the puzzle that attracts the individual, but the thought of financial gain and social status, particularly within the scientific community. Ribbon and gold, to use Chapter 2's terminology, have become more and more important.

A similar concern was echoed by the psychologist Anne Roe as early as 1952.[23] In her study *The Making of a Scientist*, she worried whether more recent cohorts of scientists were less motivated than were earlier cohorts because there were fewer obstacles, economic or otherwise, to getting through graduate school in the 1950s than there had been in earlier periods. "Now there are so many fellowships and other forms of financial aid for students who qualify on the basis of intelligence or undergraduate records that there is practically no selection on the basis of motivation. There is also some good evidence that high motivation is best maintained in the face of some, but not too many or too difficult, obstacles."[24]

One wonders whether this situation was not exacerbated in the late 1950s and 1960s as the production of Ph.D.'s boomed, fueled in part, as we have noted, by a significant expansion in graduate student support. Confirmation of this idea comes from physicists who believe that their field has actually benefited in recent years precisely because it has gone out of fashion. The anthropologist Sharon Traweek interviewed senior physicists who felt they had a stronger commitment to physics than younger physicists who came into the field primarily for "glamour and excitement."[25] Now, according to Traweek, "Some [physicists] are pleased that biology has become fashionable, because the students who follow fashion will not be in physics."[26]

The Brain Drain Hypothesis

The second hypothesis concerning cohort-related quality differences is the brain drain theory. Even if one were not worried about dipping too far into the talent pool in the United States as a result of growth, one must question the assumption that the "best and the brightest" always choose careers in science.

Interest in science clearly waxes and wanes. Many exceptionally talented individuals, for example, were drawn into physics in the early part of this century when two great revolutionary theories (the theory of relativity and quantum theory) changed the face of physics.[27] Similarly, in the 1950s and 1960s, talented persons chose careers in science in the wave of competitive excitement and patriotic enthusiasm that followed the Soviet Union's launching of *Sputnik*. Between 1956 and 1960, for example, as many as 33 percent of National Merit scholars reported that they planned careers in the physical and natural sciences. After the initial excitement and as prospects in science diminished, this percentage fell, reaching a trough of 19 percent in 1980. Since then, interest has apparently increased, with 24 percent of the National Merit scholars in 1987 planning careers in the physical and natural sciences.[28]

Not only does intrinsic interest in science change over time, but its rewards — particularly financial — also wax and wane. And scientists, or those who might otherwise be scientists, are responsive to the financial rewards.[29] Thus one reason that many talented students did not choose careers in science in the 1970s and early 1980s is that, in real terms, academic salaries throughout the 1970s and 1980s fell relative to those in the private sector.[30]

Interest in science may also change as a result of students' experiences in science classrooms as undergraduates. For example, there is a great deal of concern today over why more than 50 percent of those who enter college with a professed interest in science drop out of this field before graduation. Some — particularly faculty — believe that the dropout rate is caused by the lack of adequate precollege preparation. According to a professor of chemical engineering at Yale University, "They get here and may genuinely be interested in science, but they don't have the tools to do it." Students argue that science majors are not as attractive as others, especially given the long hours spent doing problem sets or working in labs and the "stingy" grading system that confronts them at the end of the term.[31] As a Yale undergraduate who switched from being a science major to majoring in English observed, "In other (non science) classes, if you do the work, you'll get an A. In science, it just doesn't work that way."[32]

Much of the competition for the most talented members of society today has come from professions such as business and law. And many of those who do persist in majoring in science as undergraduates have chosen to go into medicine. "Many talented young people who once might have dreamed of designing a lunar-landing module now yearn to be a different kind of "rocket scientist" — a Wall Street wizard."[33] One example is Lawrence E. Hilibrand, a high-tech bond trader at Salomon Brothers in New York. Trained in math as an undergraduate at the Massachusetts Institute of Technology, Hilibrand made headlines in 1991 when he received a bonus and salary package totaling $23 million.[34]

A less dramatic indication of the shift away from science and math by talented youth is given by the same National Merit scholar data cited earlier. They show that between 1966 and 1987, plans to major in chemistry fell by more than 68 percent, whereas plans to major in physics fell by more than 34 percent. On the other hand, the percentage of scholars choosing premed more than doubled from 1966 to 1976, peaking at 10.9 percent in 1976.[35] One has the distinct impression that there has also been a brain drain into law, although given the way that this survey defined planned areas of study, the data for law school plans are sketchy. Careers in business have also become increasingly attractive to bright students.

Other evidence concerning a brain drain is available that is more directed to the flow of exceptionally talented students who are further along in their education. Howard Bowen and Jack Schuster, for example, examined the career choices of Americans designated as Rhodes scholars over the period 1904 to 1977.[36] They found that interest in academe among this elite group began to erode in 1960. Indeed, between 1945 and 1959, almost twice as many scholars opted for careers in academics as chose careers in law, medicine, and business. But for the period 1965 to 1977 the situation was reversed, and more than twice as many scholars chose careers in law, business, and medicine than in academics.

A similar story can be told for the career choices of students elected to Phi Beta Kappa.[37] Between 1945 and 1969, slightly more than 1.2 times as many members chose careers in the professions of business, law, and medicine as chose careers in academics. But during the 1970s the situation dramatically changed, with almost five times more of this talented group entering careers in these professions than choosing academe. These results are consistent with the fact that the proportion of Harvard undergraduates graduating summa cum laude (roughly the top 5 percent of the class) who attend graduate school in the arts and sciences fell from 77 percent in 1964 to a mere 25 percent in 1981, before rebounding to 32 percent in 1987.[38]

Finally, a study of the "highest achievers" among college seniors conducted by the Consortium on Financing Higher Education (COFHE) in 1985 found that the proportion of superior students (generally those in the top 3 to 4 percent of their graduating classes) choosing to pursue doctoral studies rose from 29 percent in 1956 to 44 percent in 1966, before plummeting to 21 percent in 1976.[39] Meanwhile, the proportion opting for professional schools rose throughout the entire period, at first slowly from 33 to 37 percent between 1956 and 1966 and then more quickly, rising from 37 percent to 53 percent between 1966 and 1976. COFHE also examined graduate admission records from 1972 to 1980 at twenty research universities and found that admission during the period was offered to larger proportions of shrinking applicant pools. Again, this is evidence of a brain drain.[40]

The "brain drain" evidence is not, however, unanimous. In order to obtain information concerning the quality of graduate students by field, Bowen and Schuster surveyed the highest-ranking 15 percent of departments in each of thirty-two fields, asking chairpersons for a comparison between "current advanced graduate students" and their counterparts "during the 1968-1972" period.[41] In the physical sciences their study included chemistry and physics. For the earth sciences it included geosciences, and in the life sciences zoology, microbiology, biochemistry, and botany. With the exception of botany, Bowen and Schuster found that the chairpersons in the elite life science departments perceived their graduate students as better (approximately 5 or better on a scale on which 7 was "much better," 4 was "about the same," and 1 was "much worse.") In the geosciences, a field that gained a great deal of attention following the plate tectonic revolution in the mid-to-late 1960s, the quality reported also was higher (4.9). In chemistry and physics, Bowen and Schuster found that the chairpersons perceived the students to be closer to "about the same" (4.6 in physics, 4.7 in chemistry), although still somewhat better than in the earlier period.[42]

Bowen and Schuster interpret these results as casting doubt on the argument that there has been a brain drain in the sciences, particularly in the physical sciences. Another interpretation is possible, of course, and that is that the benchmark chosen (1968–72) for the study may represent a cohort that was not exceptionally able or motivated, precisely because inordinate growth (which was clearly occurring during this period) is accompanied by declines in average quality and motivation.[43]

Another explanation of Bowen and Schuster's findings relates to the changing composition of Ph.D. students in the United States. The brain drain argument refers implicitly to U.S. citizens. To make up for the loss in "local" talent that the brain drain has occasioned, doctoral programs have looked increasingly to foreign nationals, attracted to the United States by the opportunity for a first-rate education as well as by the possibility of finding permanent employment here after completion of their education. In the late 1960s to early 1970s (the benchmark period used by Bowen and Schuster), foreign students made up about 25 percent of Ph.D. students in the sciences, mathematics, and engineering in the United States. By the mid-1980s the comparable figure was 33 percent.[44] Most affected were the physical sciences, with an increase from 20 to 30 percent, mathematics with an increase from 21 to 44 percent, and engineering with an increase from 35 to 51 percent; least affected were the life sciences with an increase from 19 to only 21 percent. The point is that foreign students who come to the United States are generally of the highest caliber, being the best their countries have to offer. Thus, the fact that Bowen and Schuster detect little decline in quality may say less about the validity of the brain drain hypothesis than about the change in the number of foreign graduate students enrolled in U.S. science and engineering programs. As one scientist said to us, "The quality is no different, but the students sure look different."

The Sociocultural Hypothesis

The final hypothesis concerning cohort-related quality differences in the scientific community is the sociocultural hypothesis. As we noted earlier, it has two components: the relationship between relative cohort size and personal development (particularly achievement and motivation) and the relationship between the sociocultural environment and individual creativity.

Numerous aspects of personal development such as happiness, motivation, and various dimensions of adult intellectual functioning, including fluency and ability, vary by cohort over time. One factor at work is relative cohort size. As we noted earlier, the Easterlin hypothesis suggests that the social and economic fortunes of a cohort vary inversely with its relative size.

It is generally thought that crowding within three major social institutions—family, school, and labor market—can explain much of the cohort variation in the personal attributes observed. For example, because at any given time the capacity of the school system, in both physical and human dimensions, tends either to be fixed or to change in size very slowly, a surge in entrants taxes the resources of the system and results in a reduction in the quality of physical facilities and teachers per student. Thus, some would argue that the inverse correlation between standardized tests scores (such as the Scholastic Aptitude Test) and the size of the student population is not a coincidence. According to this view, test scores fell in the 1960s and 1970s because the baby boomers, who had reached the test-taking age, had had a crowded experience. But then the test scores began to rebound in the 1980s as the baby-bust cohort entered adolescence. There is, however, no hard evidence to back up this conjecture.

The second component of the sociocultural hypothesis is the contention that personal creative development is influenced by certain characteristics of the social environment. Mokyr argues that technological innovation (or for that matter, innovation of any kind) is not likely to take place in a malnourished, superstitious, or highly traditional society and that innovation requires a culturally diverse and tolerant society. Furthermore, the incentive structure offered by the society's economic and social institutions also must be conducive to innovation.[45]

In a similar vein, Simonton believes that personal creative development is influenced by factors such as the availability of role models, the extent of cultural diversity, the level of political fragmentation, and both imperial and political instability. Thus changes in these conditions over time can lead to generational fluctuations in creativity. To test this hypothesis, Simonton[46] used what he termed a "transhistorical time series analysis." That is, he explored differences in the incidence of creativity over time with a database covering acts of individual creativity, as noted in histories, anthologies, and biographical dictionaries, spanning 127 consecutive generations (twenty-year intervals) from 700 B.C. to A.D. 1849, that is, from ancient Greco-Roman to relatively modern European civilization. He found that the data were consistent with his hypothesis. Although it is not obvious that these results hold on a smaller scale than the broad sweep of history investigated by Simonton, they nevertheless give credence to the argument that generations, on average, may differ in their creative abilities.

SUMMARY AND CONCLUSIONS

We have argued in this chapter that there are three reasons to suspect that the quality of scientists varies by cohort. For convenience, we have referred to these as the selectivity, brain drain, and sociocultural hypotheses. For various reasons, all three hypotheses are somewhat difficult to test; the evidence concerning each is somewhat indirect and of the case history variety. Yet, taken together, the evidence leads us to believe that cohorts of scientists do vary in quality.

In particular, the evidence makes us believe that due to extremely rapid growth the quality of the average scientist declined in the 1960s. Then, when this growth tapered off, the quality of the average scientist remained at this lower level, as exceptionally talented students chose jobs in the more lucrative professions of business, law, and medicine. There is, however, some evidence that this pattern is reversing itself today.

In the conclusion to Chapter 6, we argued that the forces of RPRT also conspire to make certain generations, or cohorts, of scientists more productive than others are. Two factors are at play. For ease, we referred to these as job market conditions and vintage effects. The importance of job market conditions means that cohorts of scientists have had differential access to the resources that foster research. Those who have had the good fortune to receive their Ph.D.'s during a seller's market have had the greatest chance of spending their careers at a place that nurtures research. Those who have graduated when the market favored buyers have had a much harder time finding employment in the research sector. Presumably their productivity has suffered. Furthermore, we have seen that these effects, at least in regard to location, are not insignificant. Changes in the market for physical scientists have had strong repercussions on location within the scientific community. For example, in the 1950s and early 1960s the market for scientists was strong; even today, twenty to thirty years later, almost four out of ten physical scientists who got their degrees during the 1950s and early 1960s are employed at research-oriented academic institutions and FFRDCs. The market for physical scientists deteriorated in the late 1960s, however, and weakened markedly in the early 1970s. As a result, today only about one in four of the physical scientists who got their degree after 1965 still work in the research sector. The story is even more dramatic in physics than in the other physical sciences. Research resources are thus clearly linked to cohort membership.

This is not to say that scientists are not productive in business and industry, the sector that has grown in importance as an employer of Ph.D. scientists. Scientists in business and industry do engage in research. But much of this research is of a prescriptive nature, most of it is applied rather than basic, and most is designed with an eye to profit; hence the results of the research are generally not shared with colleagues working outside the company.[47] Since the late 1960s or early 1970s, however, the growth of R & D expenditures in business and industry in the United States has slackened. Thus, there is the irony that the increased employment of scientists in business and industry coincided with a decreased emphasis in business and industry on the research and development function.[48]

Job market and quality conditions taken together suggest that productivity may be particularly depressed for the cohorts of physical scientists who received their Ph.D.'s in the late 1960s and early 1970s. Not only did this group have less access to research resources than did earlier cohorts, but it also could be argued that they were, on average, of a lower quality, particularly when compared with those who received their Ph.D.'s in the 1950s. To a lesser degree, this may have also occurred in the earth sciences. There is less evidence that life scientists who received their Ph.D.'s during this time had as much trouble obtaining jobs in the research sector. Although the extremely rapid growth experienced by the life sciences may have diluted the talent pool of those going into the field, this may have been partially offset by the flow of talent within science from the physical to the life sciences. Finally, there is also the possibility that some men who pursued Ph.D. degrees in science in the late 1960s and early 1970s were motivated less, at least in part, by the love of science and more by the desire not to be drafted for service in a war that had, at best, mixed support in the United States. Once again we must caution the reader that cohort arguments refer to the average scientist in a cohort, not to any particular member of the cohort.

In addition to the market and quality considerations, the importance of vintage means that certain cohorts have been trained at a fortuitous time, receiving an education that sets the stage for a productive line of research. For them, the intellectual climate was ripe. Other cohorts may not have been so fortunate, receiving their training before a period of major change in theory or technique. As a result, their research may have suffered as they were forced to cope with the threat of knowledge obsolescence.

Not only are cohort effects important in their own right, but their presence in science also means that researchers studying the relationship between age and productivity must be careful to differentiate cohort effects from aging effects, much as the IQ researchers, whose work we summarized in Chapter 3, have had to do. The next chapter presents a study we conducted that takes into account cohort considerations when estimating the age–productivity relationship.

NOTES

1. Carlsson and Karlsson 1970, p. 710.
2. Gottlieb 1987, p. 11.
3. For an introduction to this literature, see Easterlin 1987.
4. This is referred to as the "Easterlin" hypothesis.
5. For a recent articulation of this theme by a particle physicist, see Traweek 1988, p. 100.
6. Price 1986.
7. Sovern 1989.
8. Nulty 1989.
9. See, for example, Simonton 1975, 1976, 1984.
10. Mokyr 1990.
11. The text of this classic has been included in Price 1986.
12. Price 1986, p. 17. Price exaggerated slightly, but not by much. To place his prediction in some perspective, if the number of Ph.D.'s in science and engineering were to double every ten years, by the year 2097 there would be approximately 617 million scientists.
13. See National Research Council 1978, p. 26. The estimates derived for 1920 and 1930 assume that 50 percent of the reported total population of living Ph.D.'s were in science. The estimated population of living Ph.D. scientists in 1981 and 1987 comes from National Research Council 1988.
14. Price 1986, pp. 47–48.
15. Although data collection for the DRF began only in 1946, it was possible to obtain some

basic information directly from the universities concerning the identity of Ph.D. recipients going as far back as 1920. In 1958, the National Research Council began administering the Survey of Earned Doctorates to new Ph.D. recipients, enriching the DRF database.

16. Astin 1971. The Astin selectivity index was created using data collected in the 1960s. Although the quality of colleges and universities clearly changes over time, most people would agree that the top sixty-five identified here were extremely selective throughout the period studied.

17. The data that Astin used to determine the selectivity ratings are for 1967 and 1968. Given that there is approximately a seven-year lag between the receipt of the baccalaureate and the receipt of the doctorate, this suggests that the Astin ratings are most accurate for those receiving doctorates around the period 1974 to 1975.

18. Bishop 1989.
19. Ibid., p. 195.
20. Cole and Meyer 1985.
21. Price 1986, p. 97.
22. Menard 1971, p. 5.
23. Roe 1952.
24. Ibid., p. 49.
25. Traweek 1988.

26. Ibid., p. 96. She also reports that some physicists are concerned with the large number of students switching to biology.

27. For an interesting account of this period, see Gamow 1966.

28. National Merit Scholarship Corporation. Intentions, of course, are not necessarily an accurate indicator of career outcome.

29. See, for example, Freeman 1975.

30. According to Bowen and Sosa (1989, p. 85), the actual decline in relative earnings position from 1970-1 to 1983-4 was about 17 percent. Ehrenberg and his colleagues (1991) claim that there was no drop in the starting salary of new assistant professors of physics compared with that of holders of masters of engineering degrees during the 1970-88 period. They do, however, suspect that the professions have offered many more opportunities for earnings growth over the career than has the academic world. They found, for example, that the ratio of senior faculty to assistant professor salaries is less than two, implying that in the academic sector where many scientists are employed, earnings grow at a much slower rate than they do in the professions.

31. Some suggest that both may be a carry-over from the days when, faced with long queues of students hoping to become scientists, faculty developed grading systems and teaching techniques designed to weed out the less serious.

32. See Dana Milbank, "Shortage of Scientists Approaches a Crisis as More Students Drop out of the Field," *Wall Street Journal*, September 17, 1990.

33. Nulty 1989.

34. See Randall Smith, "Roaring '90s? Here Comes Salomon's $23 Million Man," *Wall Street Journal*, January 7, 1991.

35. One reason that the medical profession became so appealing to students may have been that with the addition of the postdoctorate, it began to take approximately the same amount of time to complete graduate training as it did to complete a residency. Since 1980, however, the percentage choosing premed majors has declined, falling to 7.5 percent in 1987.

36. Bowen and Schuster 1986, p. 226.
37. Ibid.
38. Rosovsky 1990, p. 145.
39. See Bowen and Sosa 1989, p. 112.

40. What will happen in future years is uncertain, although there are already some signs that career interests are shifting once again. Since 1976, fewer National Merit scholars are choosing prelaw (political science) or premed majors, and the proportion of students choosing majors in the physical and natural sciences has increased (see National Merit Scholarship Corporation). The declining interest in medical schools has prompted one observer to comment: "To fill their classes, colleges of medicine must dig deeper into their applicant pool." Since 1974, applications to medical schools have dropped by more than one third, and the chances of acceptance rose from 34 percent

to 55 percent in 1987. Lucrative opportunities in business and computer science, rapidly rising medical tuition, and a "tarnished image" of the medical profession all are blamed for the drop in applications (see "Getting into Medical School Easier as Tuition Rises, Marcus Welby Image Fades," *Atlanta Journal Constitution*, August 14, 1988). Salaries for research scientists are even beginning to look brighter. In 1988 they rose by more than 9 percent, the largest increase in five years (see Nulty 1989).

41. Bowen and Schuster 1986.
42. Ibid., p. 215.
43. The Hartnett (1987) study on the brain drain hypothesis is not discussed here because of its data problems. The study linked Ph.D. graduates from large and prestigious programs with their college SAT scores for four intervals during the period 1966 to 1981. In physics and chemistry, the two areas of science that were examined, the verbal scores rose between 5 and 10 percent until 1976 and then fell by several points, whereas the math scores rose by a slightly smaller percentage and then stayed fairly stable. The problem with drawing conclusions from this study is that although 72 percent of the schools surveyed did respond, accurate scores were found for only 46.2 percent of this 72 percent. Thus, it is difficult to argue that the data are representative of what was happening during the period, especially because scores were collected for more persons in the later time periods, even though the number of new Ph.D.'s in physics and chemistry decreased by a significant amount during this period.
44. These percentages are based on the number of non-U.S. citizens receiving doctoral degrees in the United States reported in National Science Foundation 1990a. Because doctoral study takes several years to complete, the five-year periods 1971-5 and 1985-9 were used to calculate the percentages.
45. Mokyr 1990, pp. 11-12.
46. Simonton 1975.
47. Stephan and Levin 1990.
48. See Scherer 1986.

8
Age, Cohort, and Scientific Productivity: A Case Study

In Chapter 3 we discussed a variety of reasons, ranging from the economic to the sociological to the psychological, why a relationship may exist between age and productivity and, more precisely, why scientists may engage in less research in the later years of their careers. The evidence presented in Chapter 4 is consistent with much of this theory. There does seem to be an age–productivity relationship, and the relationship is discipline dependent. Moreover, it is more pronounced for eminent scientists than for average scientists, for whom age explains only a small amount of the variation in scientific productivity.

In Chapter 6 we examined the importance of place and time to a scientist. RPRT, as we called the concept, not only affects individual scientists but also means that certain cohorts of scientists may be more productive than other cohorts of scientists. Of particular importance is the fact that certain cohorts have limited access to the top research sector, whereas others have little problem finding jobs that facilitate their research. Also important is whether the cohort is educated at a time when major changes are occurring in the intellectual climate in the field. In Chapter 7, we extended the cohort story by including quality considerations, arguing that cohorts of scientists may differ in certain attributes related to scientific productivity, such as talent, motivation, and values.

When considering age and cohort effects, we also had the goal of convincing the reader that it is difficult to examine age and cohort effects empirically. The difficulty arises because the two effects are inextricably intertwined. For example, if one analyzes aging effects by studying a cross section of the scientific population, there is the problem, as we noted in Chapter 4, that both age and cohort vary within the sample. For simplicity, suppose science develops in such a way that the last educated are the best educated. Assuming that no other cohort effects are present, we would expect, other things being equal, later models of scientists to be more productive and earlier models to be less productive. Now, in a cross section, the young are the later models, and the old are the earlier models. Thus, if younger scientists are found to engage in significantly more research than older scientists, it is not clear whether this happens because they are younger or because they come from later cohorts with superior scientific knowledge. True aging effects have not been identified.

One way to separate aging effects from cohort effects is to follow a single cohort, such as the class of 1953, over time, using what is called a longitudinal design. But this approach gives no insight into whether cohort effects are actually present, because by design, the cohort does not vary. Yet the question of whether or not cohort effects are present is important. Furthermore, longitudinal data are contaminated by "time-period effects" that may be present if the research environment changes over time. For example, with the passage of time, resources may become more or less available, and other dimensions of competition in the field may change, altering a scientist's likelihood of being productive. Because of these time-period effects, the age profile produced from longitudinal data cannot necessarily be attributed to age. Thus, what could be called the identification problem will persist even if longitudinal data are used to follow a cohort over time.

There is, however, a method for isolating or, more correctly, identifying aging effects versus cohort effects while controlling for the presence of time-period effects. The methodology relies on "pooling" several cohorts of scientists. In this chapter we present a case study that uses this approach to take into account cohort considerations in estimating the age–productivity relationship. The empirical analysis focuses on the research productivity of scientists in six subfields of physics and earth science employed full time at prestigious doctorate-granting departments, for it is within this sector that the vast majority of research, at least in terms of journal publications, is produced. Before looking at the methodology in some detail, we shall define what we mean by "pure" aging effects.

WHAT ARE PURE AGING EFFECTS?

As we stated in Chapter 3, there are a variety of reasons that productivity and the passage of time may be related. In many instances, the age–productivity relationship hypothesized is indirect. Thus, developmental psychologists posit a relationship because taste or the need for recognition changes with the passage of time; sociologists posit a relationship because of the processes of reinforcement and cumulative advantage. In other instances, the age–productivity relationship is more direct. Economists, for example, hypothesize a relationship between age and productivity because of the finiteness of life: With each additional year that passes, the scientist has fewer years remaining to recoup the rewards of engaging in research.

In this chapter, when we speak of aging effects, we mean the latter, how the aging process directly affects productivity. Our goal is to discover whether – once one controls for other, indirect variables such as motivation, resources, and position in the scientific community – a relationship remains between a scientist's age per se and publishing productivity. Thus, compared with the age-publishing profiles studied earlier in Chapter 4, which do not separate the direct from the indirect effects of aging on productivity, the profiles presented in this chapter can be viewed as "purer." In other words, they attempt to reveal, other things being equal, the true relationship between age and publishing productivity. How successful we are depends on our ability to control for the host of factors that may be systematically related to both age and productivity.

METHODOLOGY

Before turning to our results, we shall briefly summarize the methodology used. We begin with the concept of pooling and the issue of identification within a pooled sample. Although we have tried to make our discussion as nontechnical as possible, some readers may wish to skip ahead to the presentation of the findings.

Pooling and Identification

Cross-sectional data are valuable to researchers because they control for time-period effects. By design, there is only one time period. Cross-sectional data are insufficient in many instances, however, because as we have seen, cohort and aging effects are commingled. Longitudinal data, which follow a cohort over time, eliminate the contamination of aging effects with cohort effects. But longitudinal data commingle aging effects with time-period effects and do not permit the researcher to see whether differences exist among cohorts. What is needed is a design that captures the useful aspects of both the cross-sectional and the longitudinal designs while eliminating the problems they present. One means of doing this is to pool several cross sections, say ten, by putting together data collected at ten different time periods for the same group of scientists.

The advantage of using such a design is that cohort can be held constant while observing aging effects, and age can be held constant while observing cohort differences in productivity. For example, for members of any specific cohort (say the class of 1953), there will be ten distinct observations, corresponding to ten different ages for each person. Thus, one can observe the aging effects over a ten-year period while holding the cohort constant. At the same time, for any age (say 35) there will be observations of the behavior of ten different cohorts (if cohorts are defined in terms of an annual date) at that age. Consequently, one can observe differences in productivity due to cohort while holding age constant. The quick reader will, however, note that there is still a problem. We have not controlled for time-period effects. Thus, the age effects observed by following the class of 1953 over time may be due to time-period effects rather than to true aging effects (both age and calendar year have changed), and the cohort effects observed by tracking the behavior of 35-year-olds over the ten-year interval may also be due to time-period effects rather than to true cohort effects (both cohort and calendar year have changed).

In order to rectify this problem and identify the true aging effects, we take two steps. First, we include in the analysis, in addition to a scientist's cohort and age, a variable designating the calendar year. Second, we ensure that the three variables—age, cohort, and calendar year—are not perfectly correlated, by defining one of the three variables, in this case, cohort, not in terms of a specific year but instead by grouping several years together.[1]

In our study, we pool data on scientists collected in four different waves of the Survey of Doctorate Recipients (SDR).[2] The earliest cross section was compiled in 1973, the latest in 1979.[3] The SDR, as we noted in Chapter 4, is the largest and most comprehensive longitudinal study of scientists in the United States. To measure research productivity, we count the number of articles published by each scientist in the two years immediately after the survey date. Four different measures of published articles are computed: a straight-count measure, referred to as PUB1, which includes all of a scientist's articles appearing in the given time frame, regardless of the number of coauthors or the quality of the journal in which the article appears; an author-adjusted measure, PUB2, which apportions shares of articles to an author depending on the number of coauthors; a "quality"-adjusted measure, PUB3, which weights each article by the "impact" of the journal in which the article appears; and finally, PUB4, a measure that adjusts for both the number of coauthors and the quality of the work. We discussed these measures in more detail in Chapter 4.

In order to disentangle age, cohort, and time-period effects, we combine several Ph.D. classes to form cohort groupings. Not only does this permit identification, but it also makes theoretical sense because, as we saw in Chapters 6 and 7, a scientist's productivity does not depend on the specific year the Ph.D. was received, but, rather, on conditions in the field, common to several classes. Furthermore, our investigation of cohort effects suggests that these cohort groupings must be made at the subfield level of a discipline. Thus, we made case studies of the selected subfields.

In addition to the age, calendar-time, and cohort variables, we include several other explanatory variables in the analysis. One variable measures the reputational rank of the academic department in which the scientist works and is included as a proxy for the richness of the scientist's research environment.[4] Another variable measures whether the scientist has a heavy teaching or administrative load and is included to stand for research effort. In addition, we incorporate two other variables, a self-reported measure of whether the scientist has received federal funding to do research during the study period and the scientist's salary, measured in inflation-adjusted dollars. The latter variable serves as a proxy for previous productivity.

Besides the question of identification, several other methodological issues must be addressed before we can estimate either an age–productivity or a cohort–productivity relationship. These include the presence of sample selectivity bias, the presence of unmeasurable individual-specific fixed effects such as talent and the taste for science, and the presence of a limited dependent variable.[5]

Sample-selectivity Bias

As we noted in Chapter 4 and considered again in Chapter 6, research output varies significantly across employment sectors. With few exceptions, scientists employed in doctorate-granting departments write significantly more articles than do scientists employed elsewhere.[6] This occurs partly because such departments are often rich with resources or, at least, rich with the connections that produce additional resources. In addition, a high level of collegial exchange that fosters productivity is common in such environments, and these departments attract bright, eager graduate students, who greatly facilitate research.

Because the research environment varies across employment sectors, it is important in a study of research productivity either to control for this variability or to limit the study to one particular sector. Because it is difficult to control for the variability[7] and because the vast majority of research is done at doctorate-granting departments, we limit this portion of our study to scientists employed (in their fields) in these departments, with the exception that for the subfield of particle physics, scientists employed at FFRDCs (federally funded research and development centers) are also included. Limiting the investigation to scientists employed at these elite institutions, however, introduces another problem, the problem of "sample selection." At issue is that productivity may be high at these institutions not only because of the presence of good colleagues, top graduate students, and access to ample resources but also because these institutions shop for talent. Professors at top schools tend to be above average in the ability pool of scientists. In itself, this would not present a problem for estimation. Rather, the problem arises when one considers that within the top sector there may be a relationship between age and ability. Elite universities hire relatively many young assistant professors but retain only the best. Tenure at a top university is granted to only a small percentage of the assistant professors. At Harvard, for example, in the Arts and Sciences, tenure is awarded to only about two out of ten professors.[8] Thus, age may be correlated with ability (a characteristic that we are unable to measure directly for members of the sample). As a result, the older persons in the selected sample may publish more than do the younger ones because they have more of what it takes to do science, not because they have become more productive with age. In this case, the difference between the observed age–publishing relationship and the true relationship can be attributed to what is called *sample selection bias*.

To control for this bias, we perform a two-step procedure first proposed by James Heckman in 1976.[9] We estimate for each survey year a multiple regression predicting the likelihood that a scientist is employed in the selected sector and then calculate what is referred to as a *selectivity bias–correction variable*.[10] Included in each regression, among other variables, are categorical variables to capture the differences in job market conditions experienced by various Ph.D. cohorts.[11] From 1958 onward, these job market variables were constructed using data from the Survey of Earned Doctorates[12] concerning the state of the labor market for Ph.D.'s in science. For conditions before 1958, the job market variables were constructed by relying on historical accounts of the state of scientific labor markets.

Individual-specific Fixed Effects

To create our database, we pooled four cross sections; thus, by design, we have as many as four observations on the same scientist.[13] To see why multiple observations can create a statistical problem, it is necessary to review some basic properties of multiple regression analysis.

Anytime a regression is performed, one can use the regression coefficients to predict the behavior of an individual possessing a particular set of measured characteristics. Suppose, for example, that for a scientist with a given set of values for the explanatory variables, the regression coefficients predict that three articles will be written in a given two-year period. We can then compare this predicted value with the actual number of publications that the scientist writes and compute the "error." Thus, in this example, if the scientist actually writes four articles instead of the predicted three, the error will be one (article). Now, if the regression coefficients (that is, the effects of the explanatory variables) are not to be biased, these errors (there will be one for each observation in the sample) must not be correlated with any of the explanatory variables. If they are, the regression procedure will "miscalculate" (will bias) the estimated effects of the independent variables. For example, suppose that one of the explanatory variables indicates whether or not the scientist has received external funding. Because it is quite possible that scientists who consistently publish more than the regression predicts may also consistently receive funding to do research, the regression coefficients will be biased.

Pooling, unfortunately, increases the likelihood that the errors will be systematically related to the independent variables. Scientific productivity, as we saw in Chapter 2, depends on a variety of factors. Many of these, such as the taste for recognition, "scientific smarts," and the love of the puzzle, cannot be measured (or at least not measured in traditional databases such as the one we use here). Consequently, we cannot directly enter them in the regression as independent variables. On the other hand, these characteristics are, in all likelihood, invariant for individual scientists, at least during a certain period of time. Thus, individuals who love science the first time we observe them in our database are likely to love science the next time they are observed, even though we cannot directly measure their "love of science."

Although these person-specific unmeasurable characteristics may be randomly distributed in the population of all scientists, we think this is unlikely in a sample of scientists employed in doctorate-granting departments and FFRDCs. Rather, we think that in our sample, because of these unmeasurable characteristics, the regression errors will be systematically related to some of the determinants of publishing productivity such as the scientist's success in garnering external research support. And as a result, the estimates of the coefficients of the publishing-productivity model will be biased unless some correction is made for this in the estimating procedure.

The correction we choose to make is to estimate what is called a *fixed-effects model*.[14] This entails including what is called a dummy variable for each scientist in the publishing-productivity equation. Thus, if we have a sample of eighty particle physicists each of whom was observed 4 times, for a total of 320 observations, we would include seventy-nine individual-specific dummy variables (in order to estimate the equation, one "dummy" must be left out). The inclusion of these fixed effects controls for the unmeasurable characteristics and yields (at least in an ideal world) unbiased regression coefficients.

There is, however, a major cost to the fixed-effects model: Once the fixed effects have been added, it is no longer possible to include a variable in the regression that captures cohort membership directly. By definition, the cohort of the scientist is part of the fixed effect because cohort is invariant over time. That is, James Watson will always be a 1950 biologist, regardless of when we observe him. This does not, however, cause a problem for obtaining good estimates of aging effects, since the fixed-effect variables indirectly control for the cohort.

The problem arises because we wish to know the effects of cohort, not just the effects of aging.[15] In order to address both questions, while recognizing the problems that exist, we employ two estimating strategies. The first estimating strategy, known as Model I, includes the fixed-effect variables and gives, in our opinion, the "purest" estimates of true aging effects.[16] The second, referred to as Model II, deletes the fixed effects but includes dummy variables for specific cohort–vintage effects. Since we do not control for unmeasurable individual-specific fixed effects in this specification, the inferences drawn about the vintage–cohort effects must be viewed with caution as the coefficients may be biased. They are, however, the best that can be obtained under the circumstances, and they give, in our opinion, the best empirical picture to date of a direct cohort–productivity relationship.

Limited Dependent Variable

Scientists do not write, at least in the literal sense, a negative number of articles. The fewest they can produce is zero, and a surprising number do precisely this. For example, approximately 35 percent of Ph.D. biochemists publish no articles in a two-year period; in physiology, the comparable figure is 40 percent. The situation is even more skewed in the physical sciences: More than 50 percent of Ph.D. physicists publish no articles in a two-year period, and more than 60 percent of earth scientists do not.[17]

In the presence of a limited dependent variable such as publications, which cannot take on a value less than zero, the traditional ordinary least-squares regression (OLS) analysis gives biased and inconsistent parameter estimates. In cases such as publications, in which the dependent variable has a number of its values clustered at zero, the preferred procedure, developed in 1958 by the Nobel Prize–winning economist James Tobin and eponymously called *Tobit*, is to estimate the parameters of a likelihood function that takes into consideration both the determinants of the number of (positive) publications and the probability that publishing productivity is positive.[18] It is this procedure that we use here. Simply put, the procedure allows for the fact that the model being estimated is nonlinear, particularly as it approaches the zero-article count.[19]

FINDINGS

Areas of Science Studied

Publishing equations are estimated for three areas of physics and three areas of earth science.[20] The areas were chosen because they employ a large number of scientists and are well-enough defined to permit case studies. The areas studied in physics are solid state/condensed matter, particle (high energy), and atomic and molecular. In earth science, they are geophysics, geology, and oceanography.

In lay terms, researchers in solid-state/condensed-matter physics study why substances have certain electrical properties, as well as other properties such as color and translucence. It is the largest subfield in physics and has had, particularly since the development of the transistor in 1947, a major impact on everyday life. More recently, many physicists in solid state/condensed matter have concentrated on developing commercially viable superconductors that can conduct electricity without resistance. To date, the superconductors that have been developed work only at extremely low temperatures or, in the case of "high-temperature" superconductors, have other problems, such as extreme brittleness, that severely limit their commercial success.[21]

In the late 1940s, physicists became aware that the neutrons and protons within the atomic nucleus are formed by still more elementary or fundamental particles. Scientists in elementary particle or high-energy physics, as it is sometimes called, focus on these small bits of matter, the building blocks from which everything is made. Abstract theorists working in particle physics seek the laws governing the four fundamental forces—nuclear (strong), electromagnetic, weak, and gravitational—with the hope, as Stephen Hawking says, of "finding a unified theory that will explain all four forces as different aspects of a single force."[22] As we noted earlier, this search for the "holy grail" has aspects of a pseudoreligious quest, given to particle physicists by Einstein, who searched in vain in the later years of his life for a theory that would unify the forces of electromagnetism and gravitation. Experimentalists in high-energy physics use accelerators to study particles, often looking for the footprints of the particles that theorists have predicted.[23] As the particles have become smaller, the amount of energy required to detect them has increased enormously. Both experimentalists and theorists are employed at FFRDCs such as the Stanford Linear Accelerator (SLAC), Brookhaven National Laboratory, Fermi National Accelerator Laboratory (Fermilab), and the Center for European Nuclear Research (CERN), as well as at universities.[24]

Atomic and molecular physicists study the structure of molecules and atoms. Or more precisely, they study the behavior, arrangement, motion, and energy states of the electrons that orbit the atomic nucleus. Although the fundamental equations that describe the structure and properties of atoms and molecules come from quantum mechanics and, as a result, have been known for approximately seventy years, their solutions have remained elusive to theorists because of the number of variables involved. The development of lasers in the early 1960s gave experimentalists in the field an enormous boost because of the laser's ability to measure energy levels precisely.

The three areas studied in the earth sciences are oceanography, geophysics, and geology. For the most part, the emphasis in these fields is on observation and classification. The earth sciences are considered to be substantially less codified than are disciplines such as physics and chemistry. That is, in the earth sciences, a wide range of acceptable methods and underlying theories are used to describe the phenomena observed.

Oceanography relies heavily on geophysical theory and methods to investigate the oceans and lands beneath them. Observations of the oceans' floors collected from research vessels crisscrossing the oceans as well as from deep-sea drilling projects have proved invaluable to developing the concept of the earth as a dynamic body. This concept, important to the theory of plate tectonics, revolutionized thought throughout the earth sciences in the 1960s. Briefly stated, the theory of plate tectonics describes the evolution and dynamics of the earth's crust in terms of plates that are constantly shifting over the earth's surface. In the process, continents are broken up and reassembled; mountains are built; and ocean basins are opened and closed. Today, oceanographers, as well as other earth scientists, continue to explore the workings of plate tectonics.

Geophysics uses the basic principles of physics to study the structure, make-up, and development of the earth. Data from the measurement of gravity, heat flows, and seismic activity are compared with the predictions derived from conceptual models. Geology, the broadest of the three subfields of earth science studied, focuses on the earth's formation, composition, history, and changes. In our study, geology includes the specialties of mineralogy, the study of minerals; petrology, the study of rocks; stratigraphy, the study of the layering of rocks; sedimentation, the study of the environments in which ancient sediments were deposited; paleontology, the study of fossils; structural geology (tectonics), the study of the earth's structure from the perspective of the causes, processes, and effects of rock deformation; and geomorphology or what was formerly called physiography, the study of the origin and occurrence of land forms.

Aging Results

The aging effects estimated from Model I are presented in Table 8-1 for earth scientists and in Table 8-2 for physicists. These effects are, in our opinion, the "best" estimates of true aging that can be obtained from the data and reveal

Table 8-1. Age–Publishing Profiles in Earth Science by Field and Measure of Output

	Measure of output			
	PUB1	PUB2	PUB3	PUB4
Geology (130 observations)				
Age	−0.38[b]	−0.28[b]	−0.91[c]	−0.46[b]
Shape[a]	A	A	A	A
Pseudo R^2	0.21	0.30	0.19	0.25
Geophysics (69 observations)				
Age	2.37[d]	1.22[d]	5.90[c]	3.09[c]
Age squared	−0.02[c]	−0.01[d]	−0.05	−0.03[c]
Shape[a]	B(59)	B(55)	B(58)	B(53)
Pseudo R^2	0.41	0.50	0.20	0.25
Oceanography (51 observations)				
Age	0.03[c]	0.45	−2.15[c]	−2.04[d]
Age squared	−0.02[d]	−0.01[d]		
Shape[a]	C	C	A	A
Pseudo R^2	0.38	0.40	0.15	0.20

[a] See text for explanation of shape. A one-tail test of an (eventual) age decrement is applied.
[b] Indicates statistical significance at .10.
[c] Indicates .05.
[d] Indicates .01.

what happens, on average, to productivity as a scientist ages, rather than the indirect consequences on publishing productivity of the passage of time and the operation of the reward system of science. Moreover, because of the fixed effects, these estimates control for systematic differences between older and younger scientists in such unmeasurables as scientific "smarts" and motivation and for time-period effects such as differences over time in the ease of publication.

Each table indicates whether age was found to be statistically significant in the publishing equations and, if so, the level of statistical significance. (Recall that, loosely speaking, a 5 percent level of statistical significance means that there is a 95 percent chance that the results are not due to "luck.") The tables also report the value of a "pseudo" R^2 statistic[25] for each equation. The pseudo R^2 is an intuitive measure of goodness of fit in the case of models estimated by the maximum likelihood technique such as Tobit, and it can be interpreted in the same fashion as the more common R^2 statistic can. In other words, it indicates the percentage of variation in the dependent variable that can be explained by the explanatory variables included in the estimated equation.

Most important to us, the tables tell for each field the shape of the age–publishing relationship that produced the "best fit." The letter A indicates that publishing activity falls off with age at a constant rate; a B indicates that the age–publishing profile takes the form of an inverted U; and the number following the B gives the age at which the profile peaks. The letter C shows that publishing activity falls off at an increasing rate with age, and a D indicates that (at traditional levels of statistical significance) no age decrement could be discerned.

Finally, the entries for age in each table show more precisely how age is related to the four measures of publishing activity. When the age variable is entered directly in a linear fashion, the reported coefficient shows the effect of one additional year of age on the specific measure of productivity. For example, the entry of -0.84 under particle physicists employed at FFRDCs in Table 8-2 means that publishing activity, as measured by straight counts of articles, declines on average by 0.84 articles as the particle physicist ages another year.[26] When age is entered into the analysis in a quadratic form as the variables age and age squared, a nonlinear aging effect is given that recognizes that the change in publishing activity associated with age is not constant but, instead, age dependent.[27] In order to interpret these coefficients, specific ages must be plugged into the estimated model.[28]

Table 8-2. Age–Publishing Profiles in Physics by Field and Measure of Output

	Measure of output			
	PUB1	PUB2	PUB3	PUB4
Atomic and molecular physics (77 observations)				
Age	1.34[b]	0.54	4.01[b]	−0.03
Age squared	−0.02[b]	−0.01[b]	−0.05[b]	−0.01
Shape[a]	B(40)	B(30)	B(40)	D
Pseudo R^2	0.34	0.32	0.27	0.22
Particle physics: academia (149 observations)				
Age	0.03	0.08	0.74	0.74
Shape[a]	D	D	D	D
Pseudo R^2	0.21	0.54	0.19	0.30
Particle physics: FFRDCs (117 observations)				
Age	−0.84[d]	−0.41[d]	−3.19[d]	−1.27[d]
Shape[a]	A	A	A	A
Pseudo R^2	0.32	0.38	0.14	0.21
Solid-state/condensed-matter physics (159 observations)				
Age	2.43[d]	0.91[d]	6.39[d]	2.45[c]
Age squared	−0.03[d]	−0.01[d]	−0.07[d]	−0.03[d]
Shape[a]	B(45)	B(41)	B(45)	B(40)
Pseudo R^2	0.33	0.42	0.22	0.26

[a]See text for explanation of shape. A one-tail test of an (eventual) age decrement is applied.
[b]Indicates statistical significance at .10.
[c]Indicates .05.
[d]Indicates .01.

Overall, the findings are consistent with the presence of true aging effects. In almost all instances, the Tobit age coefficients are statistically significant, often at the 1 percent level, and indicate that research output declines with age or eventually declines with age. Two general patterns emerge: those in which output declines throughout the career (shapes A and C) and those in which output initially increases with age and then eventually decreases (shape B). Tables 8-1 and 8-2 also reveal that from a qualitative standpoint — with but one exception — the shapes are not sensitive to how published articles are counted.[29] In addition, not surprisingly, the values of the pseudo R^2 statistic are often unusually high, reflecting for the most part the inclusion of the individual-specific fixed effects.

In the fields of geology and oceanography, as well as for particle physicists working at FFRDCs, we find evidence that publishing activity declines throughout a scientist's career. And these declines are not insignificant when compared with the average amount of publishing done over a two-year interval by scientists employed in the field (see Table 8-3). For example, we find that with the passage of each year, particle physicists working at FFRDCs write about 0.84 fewer articles per period (PUB1), a large decrement compared with the mean of 2.50 articles per period. (The reader is reminded that due to the linearity of the age–publishing relationship in this subfield, the youngest write significantly more than do the "average" for this field.) The ticking of the clock reduces the productivity of geologists by about .38 articles per period, again a

Table 8-3. Publishing Productivity per Two-Year Period by Field[a] and Measure of Output

	Measure of output			
Field	PUB1	PUB2	PUB3	PUB4
Earth science				
Geology				
(251 observations)	1.33	0.77	2.07	1.15
Geophysics				
(96 observations)	3.50	1.80	9.76	5.09
Oceanography				
(92 observations)	2.14	1.10	4.52	2.26
Physics				
Atomic and molecular				
(110 observations)	3.59	1.50	10.54	3.95
Particle physics: academia				
(233 observations)	2.90	1.07	10.56	3.56
Particle physics: FFRDCs				
(193 observations)	2.50	0.77	7.24	1.94
Solid state/condensed matter				
(257 observations)	3.51	1.57	9.15	4.05

[a]Includes all scientists in the selected top research sector in each field, not just those used to estimate Models I and II.

sizable decrease when compared with the average of 1.33. For fairly young oceanographers (those around 35 years of age), each additional year is accompanied by about 1.37 fewer articles; for older oceanographers (around the age of 55), the passage of a year means that they write approximately 2.17 fewer articles during this period. Both effects appear large compared with the mean of 2.14.

In atomic and molecular physics, solid-state physics, and geophysics, the age–publishing relationship is an inverted U, reminiscent of the shape for academic scientists found by Cole and by Bayer and Dutton using cross-sectional data.[30] In solid-state/condensed-matter physics, output increases with age until the early forties and thereafter decreases. For example, the total number of publications written increases by approximately .77 with each additional year for solid-state/condensed-matter physicists in their early thirties, reaching a peak at 45 and falling thereafter. A similar pattern is seen for atomic and molecular physicists, although their peak age is generally slightly younger. In the case of PUB4, however, no age–publishing relationship can be discerned. Finally, a curvilinear pattern is also evident in geophysics. Here, however, output increases over a substantial portion of the scientists' careers, not reaching a peak until their fifties. Indeed, for two of the measures the peak is closer to 60 than it is to 55.

Particle physicists employed in Ph.D.-granting departments prove the exception to the rule. For this group, we find no evidence that in a fully-specified model of publishing productivity, output declines or eventually declines with age.[31] There are two reasons that this exception is not entirely unexpected. First, more than any other group of scientists, particle physicists, and especially theoretical particle physicists, perceive themselves as an elite group whose objective is to find the "holy grail" of unification. In her book *Beamtimes and Lifetimes: The World of High Energy Physicists*, Sharon Traweek[32] states that particle physicists see their field as not only the most demanding of all areas, in and out of science, but also as being at the pinnacle of knowledge. Significantly, as Traweek points out, the fine arts and mathematics are the two areas not ranked by particle physicists in this hierarchy, primarily because particle physics is presumed to include what is best about art and mathematics.[33] This strong self-image and sense of importance gives particle physicists a unique singleness of purpose that may lead them to behave differently than other scientists do.

Second, there is a highly selective process in particle physics. Particle physicists who do not have what it takes leave the field. Backwaters appear sterile. Moreover, this selectivity is more operative in academia than at FFRDCs, where there are a significant number of jobs for "marginal" particle physicists, such as managing the production and maintenance of accelerators. Furthermore, FFRDCs, with their huge staffs, enormous budgets, and large number of postdoctorates require vast quantities of administrative effort, a perfectly respectable job for a particle physicist to assume in later years of his or her career.[34] It is thus perhaps not surprising that research activity wanes for particle physicists at FFRDCs but shows no tendency to decline with age (indeed,

Cohort Results

In Chapters 6 and 7 we described three reasons that cohorts of scientists may, at least on average, differ in their level of productivity. One argument focused on the idea that cohorts of scientists have differential access to the most productive sector, depending on conditions in the job market when the cohort comes of age. Unlike labor markets in general, which, at least since the Great Depression, have not seen double-digit unemployment, there have been (and probably will continue to be) substantial fluctuations in the market for scientists. As we have seen, much of this depends on fluctuations in the level of support that the U.S. government allocates to science. Part depends on fluctuations in the birthrate, since the demand for scientists follows to some degree the number of students of college age. These job market effects, as we have called them, have an indirect bearing on scientists' productivity, by influencing the sector in which they find employment. But once scientists secure employment in a particular sector, the job market conditions that existed at the time they sought entry into the scientific community no longer directly affect their research productivity.[35] Thus, in our analysis, we do not include variables to capture these job market effects in the publishing-productivity equations. Rather, as we noted earlier in the methodology section, we include them in the equation used to compute the sample-selection correction variable. Therefore, although their effect cannot be directly observed, they have been controlled for indirectly by the inclusion of the sample-selection correction variable in the publishing-productivity equations.

Unlike job market effects, the other two cohort effects—vintage and quality—have a direct impact on research productivity. As we stated in Chapter 7, the average level of ability and motivation, both important determinants of research productivity, are likely to vary across cohorts of scientists, according to the relative growth conditions in science and the lure of science relative to that of competing professions such as law, business, and medicine.

The vintage hypothesis discussed in Chapter 6 argues that change in scientific theories and methods threatens scientists with knowledge obsolescence. Especially vulnerable are cohorts of scientists confronted with rapid and dramatic change. Lucky vintages confront little change throughout their careers, often getting in on the ground floor when a major change is occurring; unlucky vintages are educated immediately before upheavals take place. Thus, central to the vintage hypothesis is the idea that a scientist's knowledge base is cohort related. Poor vintages can, as we contended in Chapter 6, keep abreast of change. Knowledge obsolescence does not mean that such scientists themselves are obsolete. Scientists can replace obsolete knowledge with new knowledge; and they can upgrade their stocks of human capital. But the obstacles to keeping up can be severe. Furthermore, scientists can find that keeping up takes so much time that they have little time left for research. They are, as we noted, damned if they do and damned if they do not. Moreover, both the costs of keeping current and the benefits from doing so are cohort dependent. In general, more recent cohorts are likely to incur lower learning costs, for they need less time to upgrade their stocks of human capital than earlier vintages do.

A necessary condition for testing the vintage hypothesis empirically is to determine which Ph.D. classes of scientists share a relatively homogenous knowledge base and therefore bear a similar risk of knowledge obsolescence if and when knowledge changes in their field. If a field were completely "flat," experiencing virtually no change over the years, all Ph.D. classes would be grouped together. There would be but one vintage. But if there were a major change in 1956, we would want to differentiate those scientists educated after this period from those educated before. After making these groupings and given the study design outlined earlier in this chapter, we can use the vintage groupings (denoted by categorical variables) to determine whether there are significant differences in the research activity of members of different vintages compared with a base, or reference, vintage.

Determining vintage groupings is difficult, much more difficult than determining which cohorts were at a disadvantage in getting a job in the top research sector. First, although the market effects are fairly broad, cutting across whole areas of science in a single sweep, vintage effects are far more particularistic. Innovations in theory or techniques that have a major impact on one area may have a minimal effect, at least for a significant period of time, on other areas of science. Hence, in order to study the vintage hypothesis, it is necessary to study subfields of a discipline. It was for this reason that, unlike other researchers, we chose to look at subfields of physics and earth science instead of grouping together all physicists and all earth scientists. The question remains, however, whether these subfields are "finely" enough defined to permit identifying those developments that pose the threat of knowledge obsolescence.[36] This point of view was expressed by an atomic and molecular physicist who wrote to us: "I believe that one of the problems with trying to locate revolutions in atomic and molecular physics is that the field is so diverse that nobody really knows what it is. What is a revolution to one person has never been heard of by another."

A second problem concerns how one goes about identifying vintages of scientists who share a common knowledge base and as a result are exposed to a similar amount of change. Market effects leave large footprints. The National Research Council reports the number of scientists looking for jobs; both scientific journals and the popular press diagnose the health of scientific labor markets. The footprints of vintage effects, however, are far more subtle. Only in the case of major revolutions (which, contrary to popular opinion, are fairly rare in science) is the trail highly visible. The issue of identification becomes even more murky, since it is not only change in the knowledge base that alters the fortunes of different cohorts but also change in "fashion." Just as in the world of haute couture, the world of science, as we noted in Chapter 6, has trend setters. Cohorts can wake up one day to find that they possess intellectual wardrobes that, although quite serviceable, are inadequate for solving the problems currently in style for which funding and publication space are readily available.

Finally, without the scientists' input it is difficult for social scientists such as ourselves to make the necessary cohort distinctions. Yet in our experience, scientists, particularly physicists, shy away from talking about the concept of obsolescence. Perhaps they do so because, as we mentioned in Chapter 6, the specter of obsolescence is something that scientists fear, perhaps because scientists are cognizant of all the pitfalls we just discussed, or perhaps because they cannot take time away from their research to deal with such "soft" research issues. Their reluctance to discuss obsolescence also may stem from their belief that their very training as scientists protects them against obsolescence.

For our study, vintage variables were created by identifying changes in each of the six subfields between 1933 and 1979 (the last period observed in our study) that had the potential of making scientists in the subfield obsolete. For each of the six subfields, we identified changes by performing case studies. We gathered information for the case studies through personal interviews, a small mail survey, and various publications, including those produced by outside observers of the field such as historians and sociologists of science. In interviews and the mail survey, scientists were asked to list the changes in their specialties, in either theory or research techniques, that could place Ph.D.'s trained before the innovation at a disadvantage relative to those trained afterward.

After completing the case studies, we created a set of vintage categorical or dummy variables for each subfield, to denote Ph.D. classes that received, in our estimation, a relatively homogeneous knowledge base in graduate school and thus shared a common likelihood of experiencing knowledge (or fashion) obsolescence. Thus, if the case study for the subfield suggested that a major innovation took place in 1949, another in 1955, and the last in 1967 (at least according to the cutoff date for our study), we would group scientists in this field into four categories and construct four corresponding categorical variables: $V4$ is set equal to 1 if the doctoral degree was awarded after 1967, 0 otherwise; $V3$ is set equal to 1 if the degree was awarded between 1956 and 1967, 0 otherwise; $V2$ is set equal to 1 if the degree was awarded between 1950 and 1955, 0 otherwise; and $V1$ is set equal to 1 if the degree was awarded before 1950, 0 otherwise. The earliest vintage, $V1$, was always chosen as the benchmark against which differences in productivity among vintages were measured. If the latest educated are the best educated—as a belief in the secular progress of science would suggest—then the resulting step function should rise with each successively later vintage; that is, successively later vintages should be more productive.

We examine the vintage hypothesis by seeing whether, other things being equal, publishing productivity is related to the vintage variables. As we explained earlier in this chapter, this was done after controlling for age, calendar time, and selection into the research sector, as well as for several other determinants of research productivity. For reasons already noted, however, we were unable to control for individual-specific fixed effects, such as motivation and ability, at the same time that we controlled for vintage. Thus, the vintage estimates discussed next are not as "pure" as are the aging results presented earlier in Tables 8-1 and 8-2.

Table 8-4 defines the various vintages for the three areas of physics studied (a brief description of major conceptual or methodological breakthroughs in each subfield is included) and reports their estimated Tobit coefficients. In addition, the table presents significance tests for the coefficients as well as the values of the pseudo R^2 statistic indicating what percentage of the variation in publishing productivity is explained by all the explanatory variables included in the estimated equation.[37] For each subfield, four coefficients are presented, one for each of the four measures of publishing activity used as the dependent variable.[38] The coefficient of 7.06 on the V3 variable for solid-state/condensed-matter physicists, for example, indicates that this cohort wrote approximately 7.06 more articles than did the benchmark group educated before the development of many body theory and the microscopic theory of superconductivity (BSC), as well as changes in experimental technique, were introduced. Adjustment for coauthorship (PUB2) lowers the measured difference to 2.67; adjustment for journal quality (PUB3) raises the figure to 21.26; and adjustment for both coauthorship and journal quality (PUB4) places the coefficient at about 8.5.

In a similar manner, Table 8-5 presents the vintage results for the three areas of earth science studied. In these fields because a picture emerged from the case studies of one major conceptual development—the plate tectonics revolution—coupled with gradual and continuous changes in the tools and methods of research, the potential vintage effects have been specified simply as a step function shifting in five- or ten-year intervals (depending on the sample sizes) counting backwards from the emergence (beginning in 1970) of a new generation of earth scientists schooled in the plate tectonics paradigm.

The most striking finding of Tables 8-4 and 8-5 is that at conventional levels of significance, in no field are the latest vintages more productive than the earliest, benchmark, vintage. In other words, in no instance is the coefficient positive and statistically significant for the highest dummy variable denoting the latest vintage. Stated differently, there is no evidence that the latest vintage, with supposedly the most up-to-date knowledge, engages in more research than does the earliest vintage.[39] Furthermore, in several subfields, depending on the output measure, there is some indication that the latest vintages are less productive than are the earliest vintages. This occurs in particle physics, in both academia and FFRDCs; in oceanography, for all output measures except PUB4; and in geophysics, for the fully adjusted measure of output, PUB4.

Two other findings are of particular interest. One is the fact that regardless of publication measure, all of the vintage coefficients are negative and statistically significant for particle physicists working at FFRDCs. Compared with the field-theory group, educated before 1957, later vintages write fewer articles. Furthermore, the results imply that the decrement increases with each successively later vintage, although in only a few cases is the increased decrement statistically significant.[40] The other and similar observation is that in solid state/condensed

Table 8-4. Differences in Publishing Productivity by Vintage and Measure of Output in Subfields of Physics

	Measure of output			
	PUB1	PUB2	PUB3	PUB4
Atomic and molecular physics (89 observations)				
V2: Ph.D. ≥ 1963	1.26	1.03	1.95	1.59
V1: Benchmark Ph.D. < 1963				
Pseudo R^2	0.06	0.03	0.03	0.05
V2: computers and lasers introduced as research techniques				
Particle physics: academia (168 observations)				
V4: Ph.D. > 1970	−2.69	−2.68[c]	−7.03	−9.29[b]
V3: 1964 ≤ Ph.D. ≤ 1970	−0.57	−1.22	−0.94	−3.63
V2: 1957 ≤ Ph.D. ≤ 1963	0.21	−0.76	3.60	−1.83
V1: Benchmark Ph.D. < 1957				
Pseudo R^2	0.05	0.10	0.04	0.07

V4: "redux" of field theory; birth of 20-person teams at accelerators.
V3: new physics of Gell-Mann — "the periodic table of elementary particles"; Regge poles and Venziano representations popular, field theory seen as dead end.
V2: S-matrix theory, bootstrap, and early Regge pole work in vogue, field theory in decline; birth of 5-person teams at accelerators.

	PUB1	PUB2	PUB3	PUB4
Particle physics: FFRDCs (157 observations)				
V4: Ph.D. > 1970	−8.70[b]	−3.74[c]	−31.81[b]	−10.93[c]
V3: 1964 ≤ Ph.D. ≤ 1970	−8.22[b]	−3.01[c]	−27.46[c]	−8.39[b]
V2: 1957 ≤ Ph.D. ≤ 1963	−5.85[c]	−1.99[c]	−19.55[c]	−6.08[c]
V1: Benchmark Ph.D. < 1957				
Pseudo R^2	0.03	0.04	0.01	0.02

V4: "redux" of field theory; birth of 20-person teams at accelerators.
V3: new physics of Gell-Mann — "the periodic table of elementary particles"; Regge poles and Venziano representations popular, field theory seen as dead end.
V2: S-matrix theory, bootstrap, and early Regge pole work in vogue, field theory in decline; birth of 5-person teams at accelerators.

	PUB1	PUB2	PUB3	PUB4
Solid-state/condensed-matter physics (182 observations)				
V5: Ph.D. > 1972	5.55	1.39	19.77	6.11
V4: 1963 ≤ Ph.D. ≤ 1972	6.38[a]	2.09	19.12[a]	6.85
V3: 1956 ≤ Ph.D. ≤ 1962	7.06[b]	2.67[a]	21.26[b]	8.53[b]
V2: 1948 ≤ Ph.D. ≤ 1955	7.82[c]	3.51[c]	22.73[c]	9.92[b]
V1: Benchmark Ph.D. < 1948				
Pseudo R^2	0.08	0.10	0.07	0.07

V5: renormalization introduced into solid state; research front moves into condensed-matter studies.
V4: computers and lasers introduced as research techniques.
V3: many body theory and the BSC microscopic theory of superconductivity introduced; integrated circuits widely in use.
V2: emergence of new experimental techniques such as microwaves, magnetic resonance, and neutron bombardment.

[a] Indicates statistical significance at .10.
[b] Indicates .05.
[c] Indicates .01.

Table 8-5. Differences in Publishing Productivity by Vintage and Measure of Output in Subfields of Earth Science

	Measure of output			
	PUB1	PUB2	PUB3	PUB4
Geology (172 observations)				
V6: Ph.D. > 1969	2.01	0.91	3.69	1.54
V5: 1965 ≤ Ph.D. ≤ 1969	2.68	1.14	4.58	1.70
V4: 1960 ≤ Ph.D. ≤ 1964	1.53	0.87	2.54	1.02
V3: 1955 ≤ Ph.D. ≤ 1959	1.82	0.99	2.94	1.41
V2: 1945 ≤ Ph.D. ≤ 1954	1.70	0.80	2.52	1.01
V1: Benchmark Ph.D. < 1945				
Pseudo R^2	0.13	0.14	0.11	0.12
Geophysics (78 observations)				
V5: Ph.D. > 1969	−5.39	−2.31	−20.85	−13.94[b]
V4: 1965 ≤ Ph.D. ≤ 1969	−2.43	−0.69	−11.00	−7.30
V3: 1960 ≤ Ph.D. ≤ 1964	−2.22	−0.49	−5.64	−2.95
V2: 1955 ≤ Ph.D. ≤ 1959	−2.15	−0.95	−3.09	−2.11
V1: Benchmark Ph.D. < 1955				
Pseudo R^2	0.14	0.20	0.10	0.16
Oceanography (57 observations)				
V3: Ph.D. < 1969	−3.13[c]	−1.33[b]	−5.42[a]	−3.25
V2: 1965 ≤ Ph.D. ≤ 1969	−3.13[b]	−0.70	−4.26	−2.11
V1: Benchmark Ph.D. < 1965				
Pseudo R^2	0.06	0.10	0.10	0.11

[a]Indicates statistical significance at .10.
[b]Indicates .05.
[c]Indicates .01.

matter, the results are consistent with a decline in publishing productivity among successively later vintages of physicists, although here none of the differences among successively later cohorts is statistically significant.

Overall, the vintage results suggest that vintage does matter, but not in the way predicted from a "latest educated are best educated" point of view. The most recent vintage studied, with one possible exception,[41] is never found to be significantly more productive than is the benchmark, earliest vintage and is often less productive. There are several reasons that this finding is not especially surprising. First, in doing the case studies we found that at least in some of the fields studied, more recent vintages may not have had the knowledge edge that one would presume. Part of this has to do with fashion. For example, the case study suggested that in particle physics some later vintages may have been at a disadvantage because they were trained in the S-matrix/bootstrap approach that, until recently, was thought to be a dead end.[42] Similarly, the knowledge base of an earlier vintage, trained when field theory was at its zenith, came back into style in the 1970s as unification theory began to draw on field theory.

Second, in some areas, change has not been substantial. For example, the physics case study suggested that atomic and molecular physics has not really experienced dramatic changes in thought or technique over the past forty years since the upheaval brought about by the quantum revolution. Third, in some areas, as we noted in Chapter 6, there is a role for ditchdiggers. Thus solid-state/condensed-matter physicists trained before many body theory may not have learned the new theory but may have continued to do research now considered in the backwaters. Finally, in fields such as the earth sciences that are concerned largely with observation and classification, a wide range of methods and approaches all may be acceptable to the scientific community. In these less codified sciences, therefore, as we noted earlier in Chapter 6, we doubt whether conceptual breakthroughs such as the plate-tectonic revolution and innovations in technique due to advances in computer technology would render earlier vintages less productive.

There is, however, another more speculative explanation of why the latest vintages generally proved to be no more productive than the earliest vintages. This explanation relates to the quality-cohort effects discussed in Chapter 7. During the 1960s and very early 1970s, science grew rapidly. It is possible that scientists obtaining doctorates during this period of rapid expansion may have been, on average, not as talented or motivated as were scientists coming from earlier cohorts, which represent a smaller, more elite portion of the population. As a result, even if these scientists have a knowledge edge, a "talent deficit" may make them no more productive than are their peers from earlier cohorts. The evidence presented in Chapter 7 concerning the career choices of National Merit scholars, Phi Beta Kappa recipients, and Rhodes scholars, for example, is consistent with this interpretation.

In the econometric model we estimated, a sample-selection variable was introduced to control for market conditions affecting employment location; vintage dummies, as we just reported, also were introduced to capture the possible effects of change in scientific theory and technique. No variable was introduced to control for the cohort's average ability or motivation level. Although there is a great deal of anecdotal evidence to suggest that the quality of the talent pool is cohort dependent, we could not get a quantitative fix on this dimension in our study. Because no variable directly controlled for the average quality of the cohort, it is possible that the vintage dummies reflect not only changes in knowledge but also changes in the average ability or motivation level of the cohort. This would explain why the most recent vintages are generally no more productive than the earliest vintages. Thus, even if the latest educated are the best educated, as the secular progress of knowledge would imply, the latest cohorts, on average, may not be the best if the quality of the cohort has declined. Metaphorically speaking, even if the technology available for raising wine has never been better, the average quality of the wine crop can decline if the average quality of the land used to grow wine deteriorates. The findings also suggest that this ability or motivation argument might be extended into the 1970s and 1980s, long after the growth in science peaked, perhaps because the best of the crop of new students were attracted into careers in law, business, and medicine. To continue the wine analogy, although the technology of production continued to improve, the best wine-growing land was diverted to alternative purposes, and as a result, the average quality of the wine crop continued to fall.

SUMMARY

In this chapter we explored more thoroughly the relationship between age and publishing productivity and vintage and publishing productivity for physicists and earth scientists. With respect to aging, our major finding is that with the exception of particle physicists employed in Ph.D.-granting departments, true aging effects are present in a fully specified model of publishing productivity that, among other things, controls for individual fixed effects such as motivation and ability. Stated differently, there is evidence that on average, scientists produce less output as they age. And the aging effect that we found can be attributed to age per se and not to the fact that for some reason older scientists in the sample have different attributes, different values, or differential access to resources than do younger members of the sample. Thus, when all is said and done, age does matter.

With respect to the relationship between vintage and productivity, our results, although more tentative, also suggest that for the most part, vintage does matter, but not in the way predicted from a "latest educated are best educated" point of view. With one possible exception, we never found more recent vintages to be significantly more productive than earlier vintages were. In retrospect, this outcome is not especially surprising given, as we noted, that more recent vintages may not have had a knowledge edge, in at least some of the fields.

There is, as we mentioned, another more speculative explanation of why the latest vintages proved to be no more productive than the earlier vintages, and this relates to the "quality effects" described in Chapter 7. It is possible that scientists who obtained doctorates in the 1960s and early 1970s, when science was growing rapidly, may not be, on average, as talented or motivated as are those coming from earlier cohorts that represent a smaller, more elite segment of the population. Any knowledge edge that later cohorts may possess therefore may be offset by a deficit in quality.

NOTES

1. If the three variables are perfectly correlated, the model cannot be estimated. The easiest way to see the perfect correlation is to define model in terms of birth year. Let A represent age, C represent cohort, and T represent calendar year. Once we know two of these variables, we will know the third. For example, if we observe an individual in 1990 (T = 1990) who is 27 years old (A = 27), we will know that cohort (birth year) equals 1963 (C = T − A); or if we observe an individual who is 45 years old and has a model of 1945, we will know that the observation was from 1990. The issue of identification is discussed in more detail in Levin and Stephan 1991.

2. The SDR is administered by the National Research Council.

3. The SDR is a biennial survey. The study is limited to data collected in the 1970s because at the time that funding was received for this study, 1979 was the last survey wave that could be matched with the *Science Citation Index*, the source used to collect information on the publishing activity of individuals in the sample (see Chapter 4).

4. The reputational rank variable is taken from Jones, Lindzey, and Coggeshall 1982. The other variables are derived from data in the SDR file.

5. A more technical and comprehensive discussion of these issues can be found in Stephan and Levin 1987 and Levin and Stephan 1991.

6. Strictly speaking, it is the institution and not the department that awards the Ph.D.

7. Simply including categorical variables to capture the differences among sectors will not control for the variability if unmeasurable factors such as a scientist's talent or motivation affect both the scientist's location and productivity within the scientific community.

8. See Rosovsky 1990, p. 172.

9. Heckman 1976.

10. Actually, we used the Olsen (1980) ordinary least-squares technique instead of the Heckman procedure to obtain the sample-selection correction variable, as it was less cumbersome. The selected sector in academe was defined as employment in a program in the general field of training that was rated by the American Council of Education. Thus, for example, solid state physicists employed in rated physics programs were included.

11. To make this model tractable (see the discussion of multiple criteria for selectivity in Maddala 1983, pp. 278–83), we make the following assumptions: First, Ph.D. scientists desire the best academic jobs (see, for example, Alpher and colleagues 1979; Porter 1979a, 1979b), and within academia, for the most part, mobility is downward from the more prestigious to the less prestigious institutions. Thus, to be in the selected sample, such persons must have been chosen by these elite institutions and must have met the unwritten standards for continued employment, standards that are expected to tighten or loosen according to the state of the academic job market. Consequently, we estimated by ordinary least-squares, for each survey year, the probability of sample inclusion in these Ph.D.-granting departments using the following regressors: the quality of graduate training, age, age squared, whether the respondent was born in the South, whether the respondent was born in the non-South or Canada, the age at the time of obtaining the Ph.D., job market–determined cohort effects (dummy variables), and interactions between the quality of graduate training and the market cohort effects. All of these variables, except for the categorical variables representing the quality of graduate training and the dummy variables for the cohort effects, are taken from the SDR. Data on the rankings of graduate departments over time (Cartter 1966; Jones, Lindzey, and Coggeshall 1982; Keniston 1959; Roose and Anderson 1970) are used to sort the Ph.D.-granting institutions into five categories, ranging from departments that were not ranked to departments that were in the top five.

12. These are the "percent-seeking data."

13. We always have at least two observations, and in some instances, we have three, in others, four. Note that this means that we do not have a balanced design. Although an unbalanced design necessitates an adjustment in estimating variance components, no adjustment is needed in the fixed-effects specification used here.

14. For a technical discussion of this model, see Judge et al. 1980. Although we focus here on the inclusion of the individual-specific dummy variables, remember that the model also contains dummy variables for the time-period effects discussed earlier.

15. Heckman and MaCurdy (1980, p. 56) suggest that it would be possible to retrieve separate estimates indirectly for the cohort effects after controlling for all the individual fixed effects by regressing the estimated fixed effects on cohort. We could not do so successfully because there are too few observations per person. It also is not possible to obtain separate estimates directly for the vintage effects while controlling for the individual-specific effects by dropping the time-period dummies. Because the vintage of the scientist does not change over time, no vintage estimates can be obtained when individual dummies for the fixed effects are included in the model. Finally, we note that Hausman and Taylor (1981) offer an alternative instrumental-variables technique by which it might be possible to estimate the time-varying and time-invariant determinants of publish-

ing productivity while controlling for the individual fixed effects. Given the time limitations of our data, however, we did not use their approach.

16. This model provides estimates of what is called the *within estimator*, because it uses information on only the variation within groups (individuals). In this model, the "between group" variation is captured by the differential intercepts estimated with the individual-specific dummy variables. Strictly speaking, consistency is a property of large samples and would require a longer window of time than the four periods we have. As Maddala (1987) points out, achieving consistency is not possible in most empirical work. If one failed to control for these individual fixed effects, however, we suspect that the age coefficient in the pooled model would likely reflect differences in productivity between older and younger scientists rather than differences in productivity as the average scientist aged, that is, the true effect, since the "between" variation in publishing productivity is much larger than the "within" variation.

17. See Stephan and Levin 1987. These percentages are calculated across all employment sectors. The comparable figures for scientists employed in the top academic sector are lower.

18. The goal is to maximize the value of this likelihood function, that is, to find the values of the parameters that best explain what has actually been observed. The parameter estimates that do so are called the *maximum likelihood estimates* for the model.

19. See Maddala 1983 and Killingsworth 1983 for technical discussions. Note that it is not possible with the Tobit specification to estimate the individual-specific fixed effect for a scientist who never publishes over the entire period surveyed (see Maddala 1987). Consequently, we dropped a few cases from our analysis. As a result, the sample size for estimating Model I is smaller than that used for estimating Model II. The sample used to estimate Model I does, however, still contain scientists who publish nothing in some periods, just not in all periods.

20. Scientists are classified according to their field of training as indicated either at the time their doctorates were earned (and recorded in the DRF, or Doctorate Records File), or when they were surveyed by the SDR. To allow for the fact that degree designations have changed over time, scientists were also included in a specific field, such as elementary particle physics if they reported being trained in general physics but then became employed in the field of elementary particle physics. The SDR also reports the current field of employment for each scientist.

21. See Hazen 1988, p. 266.

22. Hawking 1990, p. 69.

23. For example, Carlo Rubbia of Harvard University and the Center for European Nuclear Research (CERN) and Simon van der Meer, a CERN engineer, were together awarded the Nobel Prize in physics in 1984 for their discovery of the charged "w particle," which had been predicted originally in the late 1960s by the Glashow–Weinberg–Salam unified theory of weak and electromagnetic interactions.

24. For a description of the work environment at an accelerator, see Traweek 1988.

25. There is no unique measure of the pseudo R^2 statistic. We chose that proposed by McFadden 1974, but also see Amemiya 1981 and Maddala 1983, pp. 37–41.

26. The reader is reminded that the decline represents the flow of research in progress and is measured by the number of articles published in the next two-year period (see Chapter 4).

27. If the coefficient on the age term is positive and that on the age-squared term is negative, the age–productivity profile fitted in a quadratic specification is an inverted U. The top of the U tells the age at which productivity peaks. Until that point, productivity increases with age at a decreasing rate, and after that point, productivity decreases with age at an increasing rate. Of course, if the peak comes at a ridiculously early age (such as 10), the implication is that throughout the scientist's career, output will decrease at an increasing rate; if the peak comes at an extremely late age (such as 75), the implication is that throughout the scientist's career, output will increase at a decreasing rate. Note that the age at which the profile peaks, as reported in Tables 8-1 and 8-2, was obtained prior to rounding.

28. For example, in Table 8-2 for atomic and molecular physicists, the coefficients for the age and age-squared variables mean that other things being equal, $PUB1 = 1.34*Age - 0.02*Age^2$. The effect on productivity of aging an additional year is revealed by the first derivative of this equation: $1.34 - 0.04*Age$. Thus, at age 45, publishing productivity is falling at a rate of .46 articles, on average, as a scientist ages.

29. The level of publishing productivity implied by the model estimated depends on whose individual-specific fixed effect is considered. Thus, an analysis comparable to that performed in Chapter 4, in which a particular age-group's productivity is expressed as a percentage of the productivity of the group with the highest productivity, is meaningless. What is invariant in this model, once all the explanatory variables have been considered, is the shape of the age–publishing relationship estimated, not its position.

30. See Bayer and Dutton 1977 and S. Cole 1979. Both studies are summarized in Chapter 4.

31. In Table 8-2 the coefficients on PUB1, PUB2, and PUB3 for particle physicists in academe are not statistically different from zero using either a one- or a two-tail test. The coefficient on PUB4, however, is statistically significant using a two-tail test and is consistent with the hypothesis that for this one measure output increases with age.

32. Traweek 1988.

33. Ibid., p. 79.

34. Although we controlled for whether the scientist's primary work activity was in administration, we did not control for secondary or other administrative effort.

35. Contemporaneous market conditions as reflected by whether the scientist has government research support is taken into account in the models estimating publishing productivity.

36. A related issue is how fast the new theory spreads, even within a particular specialty area, from researchers at the frontier to those at the periphery.

37. Compared with the earlier results, the pseudo R^2 statistics reported here are lower because these equations omit the fixed effects.

38. Note that the sample sizes used for the vintage equations are slightly larger than those used for the fixed-effects equations presented in Tables 8-1 and 8-2. This occurs because the fixed-effect model could be estimated only for persons who had something published during this time period. This restriction was not necessary when the fixed effects were dropped.

39. The reader is reminded that age was controlled for in the equations.

40. At the 5 percent level, the difference between V4 and V2 is statistically significant in the PUB2 publishing equation, and at the 10 percent level, the differences between V3 and V2 in the PUB2 equation and V4 and V2 in the PUB4 equation are statistically significant.

41. Geology may be the one exception. In this field, we found that if the model were reestimated using the smaller sample (sample two) that was used to obtain the pure aging effects reported in Tables 8-1 and 8-2, for PUB1, PUB2, and PUB3, the most recent vintages tended to be more productive than the earliest, benchmark, vintage was.

42. Recent work on total unification draws on string theory, which owes some of its intellectual roots to S-matrix theory.

9
Conclusion

Is there a right age for science? Is RPRT (right place, right time) important to science? If so, what are the consequences for science and the larger community? These are the questions we address in this book. To facilitate analyzing them, we developed a framework for examining how and why science is done. We also explored theoretical reasons that age and RPRT may affect productivity. Then we reported findings from our Productivity of Scientists Research (PSR) study. In this final chapter, we restate our major conclusions and take stock of the scientific enterprise. We also speculate about the consequences that increased competition has for science and propose elements of a rational science policy.

Age

Our research suggests two conclusions with regard to age. First, as they age, scientists eventually write less. This is true for not only gross counts of articles but also counts adjusted for the number of coauthors as well as journal "impact," a crude way to control for quality differences. The data suggest that the age–publishing relationship is stronger in the physical and earth sciences than in the two life sciences that we analyzed.

In our case study of selected fields in physics and the earth sciences, the age–publishing relationship persists with only one exception after the data are analyzed in a richer manner that takes into account the presence of certain generational differences and other factors that may systematically affect publishing activity. Like other researchers, however, we have found that age explains little of the variation. We conclude that for "average" Ph.D. scientists, age does matter but it does not matter a great deal.

Second, we conclude that there is a stronger relationship between age and the ability to do path-breaking work. When we study contributions instead of scientists and limit these contributions to those for which a Nobel Prize has been awarded, for example, we find in the physical sciences that almost 75

Conclusion

percent of the winners did their work for the prize by the age of 40 and virtually none did their work after the age of 55. In physiology and medicine the relationship also exists, but here, where clinical work is more important, the relationship is less pronounced. We conclude that exceptional contributions in science, particularly in the physical sciences, are most likely to be made by scientists under the age of 40. We also conclude that the importance of extreme youth to discovery has been exaggerated.

Not only does our work support these two conclusions with regard to age and productivity, but the work of other researchers also is generally consistent with these findings. Although many of these studies are not new, it is not uncommon for researchers to say that the age-productivity results they and others have found are equivocal. Such a statement, we would argue, is not consistent with the facts. Weak, yes, especially for average scientists, but not equivocal. In most instances, scientists eventually do publish less with age. Those who contend that this is not the case do so, we suspect, out of a desire to demonstrate that older persons can still do science. For those we have called journeymen scientists, we do not take issue with this ability to do science. But we believe that the incentives to do research, as well as the forces of cumulative advantage, conspire to make research less attractive to older journeymen scientists. In the case of eminent scientists, those with the sacred spark, at some point age becomes a detriment to the production of path-breaking work, especially in the physical sciences. The science game can be played by those of all ages, but the winner—particularly when the stakes are high—is more likely to be young than old.

Taken together, the first two conclusions lead us to believe that other things being equal, the (older) scientific community of today is less productive than was the (younger) community of a decade or two earlier. It is not only, as we pointed out in the introduction to this book, that the average scientist today is several years older than the average scientist was in the 1960s and early 1970s. For average scientists, a few years do not make that large a difference. Of more importance is the fact that substantially fewer scientists are under the age of 40 today than in the past. Indeed, in the academic sector, the sector in which the preponderance of basic research is done, only 28 percent of the scientists are under the age of 40, whereas in the early 1970s the number approached 50 percent.[1]

Not only are older scientists less productive, especially with regard to breakthrough research. Older scientists may also stifle the creativity and productivity of the relatively fewer younger scientists who are working today. Such a view is consistent with Max Planck's belief that "a new scientific truth does not triumph by convincing its opponents and making them see the light, but rather because its opponents eventually die, and a new generation grows up that is familiar with it."[2] It is also somewhat consistent with the evidence collected in various case studies, although it is clear that the interval between the introduction of a new idea and its adoption is not nearly so long as would be required for a whole generation to fade from positions of authority. The more deleterious effect that an older community may have on younger scientists is not to spurn its work but, rather, because of the increased competition caused by the age structure, to encourage young scientists to pursue safe, somewhat risk-free lines of investigation, the kind for which tenure is awarded but which is unlikely to lead to revolutionary ideas. We shall return to these consequences later.

Right Place, Right Time

The effects of age are not the only ones that interest us. We are also interested in the effects of what we have called RPRT: being at the right place at the right time. A third conclusion of this book is that although it is difficult to investigate empirically, RPRT clearly matters in science. As Paul Chu, who scored a significant breakthrough in the field of high-temperature superconductivity in 1987, said, "Luck is one part of the game. You have to be at [the] right place, with the right people, in the right time."[3]

RPRT affects a scientist's career—and the scientist's productivity—in two ways. First, RPRT affects where a scientist works and ultimately the scientist's productivity, because place matters to research. It matters in terms of colleagues, access to resources, and the probability that the research work will be recognized. Access to the research sector, however, is not evenly distributed across generations of scientists but depends instead on what is happening in the job market at the time the scientist's training is completed. Some scientists have the good fortune to complete their training at a time when jobs in the top research sector abound and when science is growing. Other scientists have the bad fortune to graduate at the wrong time and consequently do not get to a place that fosters research. They lack RPRT.

Job market conditions are not the only thing that affects RPRT. RPRT also hinges on the intellectual conditions existing at the time the scientist was educated. Some scientists have the good fortune to be educated at a time and at a place where major changes in science are developing. This certainly was the case for Robert Noyce, the inventor of the integrated circuit. And it also was the case for geophysicists trained at Princeton in the early 1960s and biochemists and other life scientists trained in Julius Axelrod's laboratory at NIH in the 1960s.

Scientists trained when change is in the air, especially those fortunate enough to be trained at centers where the change is occurring, become in some sense part of the change. They are the persons who usher in the "new" science and the "new" techniques. It is not necessary that they learn the new science entirely at the time of their training. What is important is that they become keyed into new ways of thinking, much as the first persons in the 1950s who began to do research with computers became keyed into the importance of becoming computer literate. In contrast, scientists who do not receive their training at such a time or in such a place must contend with much more change. Their productivity suffers. If they choose to keep up, they will have less time for research, and if they do not keep up, they may produce research that is of little interest to the larger community. They are in a sense damned if they do and damned if they do not.

RPRT has consequences for individual scientists and for generations of scientists, since many of the conditions that lead to RPRT are generation specific. For example, certain generations of scientists have the good fortune to complete their training at a time when jobs in the research sector are readily available. Other generations are less fortunate. Certain generations are educated at a time when changes in research techniques or theory have just been, or are just being, introduced. They constitute a good vintage—harvested, if you will, at a time when conditions for intellectual change and growth are ripe. Other generations have the bad fortune to be educated at a time before major changes occur. They constitute a poor vintage, harvested at a time when conditions are less fortuitous.

Empirically, it is more difficult to track the effects of RPRT than the effects of age, especially the effects of vintage. The job sector effects of RPRT are a bit easier to stalk, especially given the collapse of the academic job market in the early 1970s, which sent numerous scientists to work outside the major research sector. With regard to vintage, in our own work we found that obsolescence, and the threat of obsolescence, is something that scientists fear but, much like death, often deny as a possibility. When we tried to differentiate "good" vintages from "poor" vintages in our case study, we were thwarted by the difficulty of defining the fields of investigation narrowly enough to test our theory rigorously. Even within subfields of physics, such as solid-state/condensed-matter or experimental particle physics, the number of avenues for research is so vast that what threatens one scientist with obsolescence may be of little consequence for another. Keeping these problems in mind, our case study suggests that at least from the perspective of the 1970s, the latest vintages are, with one possible exception, not the most productive. Such a finding challenges the assumption that the latest educated are the best educated and that vintages of scientists progressively improve over time. It is, however, consistent with the presence of fads in science. It is also consistent with the idea that later generations of scientists may be less productive than earlier generations because of a decline in the quality of those going into science.

Quality

A fourth conclusion of this book is that the average quality of U.S. scientists—particularly scientists born in the United States—has fallen during the past twenty to twenty-five years. Two factors have contributed to this decline. First, during the period of rapid growth that science experienced in the 1960s, science became less selective. If it had not done so, the high growth rate could not have been achieved. As a result, the average quality declined; the cream content of science was diluted. Second, these quality effects did not self-correct in the period of slowed growth that began in the 1970s, precisely because the slowed growth was accompanied by dramatic cuts in support for graduate students as well as by dismal employment prospects upon completion of training. Many of the best and the brightest therefore deserted the science track for careers in business and law; the cream went elsewhere.

Cohort Differences

A fifth major conclusion of our book is that job market conditions and quality considerations have conspired to make those who entered science in the late 1960s, 1970s, and 1980s less productive than were their counterparts who became scientists at an earlier time. The reasons are twofold. First, and particularly in the physical sciences, scientists who completed their training in the mid-1960s and throughout the 1970s found it exceptionally difficult to get jobs in the research sector. Many of these scientists took jobs in business and industry at precisely the time that research and development funds were being cut in this sector. In physics, for example, only 17 percent of those who were trained during this period hold jobs today in what could be called a research institution. In contrast, almost 50 percent of physicists educated in the 1950s and early 1960s hold jobs today in this sector. A second reason that the entrants of the past twenty to twenty-five years may not be as productive relates to the quality considerations just enumerated. Initially, rapid growth in science necessitated digging deeper into the talent pool; then the best were attracted to nonscience careers as the relative rewards to science fell.

Increased Competition

It is not only age, quality, and job market conditions that have conspired to make today's scientific community less productive than it otherwise might be. A sixth conclusion of this book is that increased competition also has stymied creativity and productivity in science.

Science has always been competitive. The metaphor of the game is not new, and competition is arguably good for science, at least up to a point, because the desire to win is an important motivating force in science. But unlike the economy, in science it is not always beneficial to increase the level of competition. Here we argue that during the past twenty years the competition for both jobs and grants has become so intense that it has led scientists to behave, as we shall see, in what could be described as a dysfunctional manner. That is, science has become too competitive.

The phenomenon of an older scientific community, coupled with a slowdown in the growth of federal funding and a leveling off in college enrollments, has created increased competition in science for jobs, particularly jobs in the top research sector. Increased competition has meant that institutions demand more from young researchers seeking tenure and more from applicants seeking jobs. The crossbar on the tenure hurdle has been raised, at least quantitatively, and the expectations of what new applicants should bring to the job have also become higher. In the past, institutions were content to look for promise and to assess that promise in terms of where and with whom a scientist had trained. Today, institutions seek individuals who have already demonstrated promise, in terms of both published articles and, perhaps more importantly, their ability to obtain funding. In some fields it is difficult to obtain a research position without having funding in hand.

Conclusion

The competition for funding is felt not only by scientists starting their careers but by those at all levels. Whereas half of all grants deemed worthy were once funded, a third or less are funded today.[4] The competition is so fierce that the funding process is often described as a sweepstakes. "If you happen to be one of the lucky few, you get hundreds of thousands of dollars [over the life of the grant]. But otherwise you get nothing, and there's nothing to tide you over until the next round."[5] The sweepstakes analogy is consistent, too, with the fact that it can be difficult to differentiate fundable projects from nonfundable ones. As an NSF spokesperson said, with regard to which of fourteen centers to fund, "It's no longer a matter of scientific merit. You've got to split hairs."[6]

The reason for this increased competition is clear: Both the applicant pool and the size of the funding requests have grown at a faster pace than the funding pool has, particularly the pool available for "small" projects, or what are sometimes referred to as SIPs (single-investigator projects).

During the late 1950s and early 1960s a whole generation of scientists was raised on the idea that funds in science were readily available. And they were. The "while you're up get me a grant" mentality was pervasive precisely because it was descriptive. There were years in science when research funds, in real terms, grew at a rate of 15 to 20 percent. One thing this rapid growth did was convince university administrators of the ready availability of funds for research. Everything the scientist did, from making phone calls and using pencils to paying publication costs in journals, became reimbursable from grants. It was not only established research institutions that began to understand and subscribe to the importance of grants. In many instances, new institutions, as well as second- and third-tier institutions, began to see grants as the road to institutional salvation, particularly when faced with dwindling support from state legislatures and private contributors. Consequently, today not only scientists but also employing institutions see grants as inordinately important. Some argue that the emphasis on grants comes more from the institution than from the scientist. "Most universities push their faculty members to apply for grants rather than work on projects of their own and their students' interest."[7] Any research institution worth its salt has a vice-president for research, a euphemism for vice-president for external funding. Such offices not only facilitate grants, they also are part of the institutional culture that pressures scientists to seek funding aggressively.[8]

This increased pressure to seek funding came at just the time that the pool of federal funds for university research began to grow at a slower rate. Indeed, in real terms the level of funding available for research actually decreased in the late 1960s and early 1970s and then remained fairly flat for several years.[9] This was followed by a modest increase that faded when real funding levels again fell in 1981 and 1982. Since 1983, the trend in federal funding has been upward, although the increases are small compared with the growth rates of the late 1950s and early 1960s.

While the rate of growth in funding has slowed, the cost of doing research is increasing, in many instances at a faster rate than that of inflation. Several factors have contributed to this. First, and perhaps most important, indirect costs (the amount that a university or institution charges in order to recover the costs of using its infrastructure—existing plant and equipment) have gone up.[10] Second, increasing numbers of investigators and rising salaries (and the fringe benefits that accompany them) have driven up the personnel component of direct costs.[11] Third, what could be called "sophistication inflation" has made equipment more expensive.[12] In some fields of science many of the questions that remain to be answered require more complex equipment than was needed in the past. Fourth, the funding of "big," capital-intensive projects such as the Hubble telescope and the manned space station have eaten up large chunks of science budgets.[13]

The Consequences of Too Much Competition

Increased competition for both jobs and funding has adverse effects on science. It encourages paper inflation and misconduct and takes much of the play out of science. It also discourages "scientists' willingness and ability to undertake risky projects."[14] Today in science, depending on the field, it is not uncommon for persons to have published more than fifty papers by the time they are considered for promotion and tenure. Such was not always the case. When James Watson was promoted to associate professor at Harvard in 1958 he had had eighteen papers published. (One of them, published five years earlier, described the structure of DNA.) The relatively few papers that Watson had says as much about the times as it does about Watson.[15] The greater competition, the "immense pressures by university authorities and by funding administrators for proof that one is making progress,"[16] has deleterious effects: It encourages scientists to publish more (shorter) papers or to publish what is sometimes referred to as LPUs—least publishable units. (Rather than publishing one paper, a researcher splits the paper into smaller segments.) Quality is traded for quantity.

> Most publications have gotten shorter; the number of experiments that is contained within them is far less; the amount of information is almost minuscule; and there's [sic] very great tendencies of people not having done quite the right thing: not used large enough samples, not worried about some of the variables that are responsible for the results that they get.[17]

Fragmentation of results creates a number of problems for science, not the least of which is scientists' increased difficulty in keeping up because of the sheer growth in the number of scientific papers, as well as difficulty in critically evaluating the work of peers. The "thinness" of articles also means that "almost anything that is reported now has to be redone by others several times before it's accepted as the truth."[18]

A related issue is risk. If promotion and grant review boards can count (and there is every indication that this may be what they do best), scientists are encouraged to choose research agendas that are safe (and generate a large number of research articles) rather than more promising but risky lines of

Conclusion

investigation. Funding is a major issue affecting the degree of risk a scientist assumes, because scientists who pursue risky projects and come up short are unlikely to get their grants renewed. An assistant professor of biology at the University of California at Berkeley stated that after failing to get a grant renewal, "I have [since] become more conservative in my research, rather than taking risks on potentially exciting areas."[19] The problem is not just one of grant renewal. There seems to be a bias in granting institutions against funding risky research.[20] This aversion to risk, and a taste for what Robert Ballard—the director of the Center for Marine Biology at Woods Hole—describes as "mediocre everything,"[21] can even encourage scientists to hold back from a granting agency discoveries that appear too spectacular.[22]

Another consequence of the greater competition in science today is the presence of misconduct and fraud. Although not new to science (the "Piltdown man" hoax and Cyril Burt's manufactured twin data are well-known cases of fraud[23]), there is widespread concern that fraud and misconduct are growing with the pressure on researchers to produce. The importance of getting the "right" answers can lead researchers to engage in a variety of actions that are less than ethical. At one extreme—and probably the least pervasive—is outright cheating or the fabrication of data. One of the most celebrated cases in recent years of such an instance involved the Harvard cardiologist John Darsee, who in 1983 was debarred by NIH for fabricating data.[24] More recently, the case of Thereza Imanishi-Kari and her coauthor David Baltimore, a Nobel laureate and former president of Rockefeller University, has gained national attention.[25] The case has also called into question the long standing belief that the scientific community can police itself, ferreting out scientific misconduct and fraud. After lengthy investigations, it was discovered that Imanishi-Kari had committed fraud in producing the paper in question and later faking data in her laboratory notebook to provide after-the-fact evidence to support statements made in the paper. Although Baltimore had no part in the deception, he was responsible for validating the paper's authenticity because he was the senior author. Yet he failed to do so and for five years strongly defended the paper, regaling Congress and the Office of Scientific Integrity of the National Institutes of Health for inappropriately prying into the normal workings of science.

The Baltimore case suggests another byproduct of greater competition in science: The need to publish in quantity encourages gratuitous authorship. That is, a prominent scientist who has contributed little to the research may be invited to be listed as a coauthor. However, the coauthor may not scrutinize the work to see if errors or misconduct have occurred.

Sloppiness is another manifestation of misconduct in science. The pressure to publish, to get results fast, causes researchers to document findings in a hasty manner. Experimental data may not be recorded, or perhaps not recorded until a later date. Data may also not be retained, thereby denying future researchers the opportunity of reexamining them.[26] For example, it now appears that in addition to prematurely staking their claim to the discovery of nuclear fusion at room temperature and in the process bypassing the normal peer review system, the Pons-Fleischmann cold fusion team at the University of Utah carelessly performed their experiment, with the result that the experiment cannot be replicated by others. Moreover, they compounded their error by being somewhat hostile and uncommunicative to other scientists attempting to replicate their findings.[27]

A more subtle consequence of increased competition is a loss of fun. Science has been characterized as play behavior carried into adulthood. In recent years, however, such competition has meant that science has become more work and less play. The funding crisis, as some would call it, has led scientists to pursue research agendas for which funding is available, not necessarily the research that most interests them. The crisis in funding also requires that scientists spend many hours writing research proposals and submitting reports to granting agencies. Paper inflation, and the mediocrity that may accompany it, has the same result. This loss of play has taken its toll on science. The Nobel laureate Richard Feynman tells how he overcame his feelings of being professionally burned out when he decided "to play with physics." Within a week he observed, while in the Cornell cafeteria, that the medallion on a plate went around faster than the wobble of the plate. With this observation, the fun returned for Feynman, and with fun came his all-important contribution to field theory.

> And before I knew it (it was a very short time) I was "playing" — working, really — with the same old problem that I loved so much. . . . There was no importance to what I was doing, but ultimately there was. The diagrams and the whole business that I got the Nobel Prize for came from that piddling around with the wobbling plate.[28]

A final consequence of increased competition is to make careers in science less attractive to promising students. Many students are attracted to science precisely because it is fun. The drive to solve the puzzle is a major motivator. But when students see only work and no play in graduate school, when established researchers must give up their Christmas Eves to write grant proposals, discouragement sets in. Some of the best and the brightest may be lost to science.

The Long-run Consequences of Impaired Productivity

Here we have argued that four factors have led the U.S. scientific community to be less productive today than it otherwise might have been. The four, which are interrelated, are (1) an increase in age, (2) a decrease in quality, (3) a decrease in the proportion of younger scientists working at research institutions, and (4) an increase in competition.

The productivity consequences are of concern not only to scientists and science but also to the nation. Science affects our standard of living. Indeed, much of the growth experienced by our economy in the latter part of the twentieth century can be attributed to advances in scientific knowledge.[29] To science we owe a longer life expectancy, electronic innovations, and the hope for a better future.

The lag between discovery and economic growth, however, is long — around ten to thirty years in some fields.[30] Thus, the consequences of an increase in age, a decrease in quality, a decrease in the proportion of young scientists at research institutions, and an increase in competition will be felt for years to come. The piper has just begun to be paid.

Stop-and-Go Funding

The irony is that to a large extent these conditions have been caused by fluctuations in national policy, or what some would call a stop-and-go funding policy. The late 1950s and 1960s witnessed a golden age of science. Funding grew — in some years at rates higher than 15 percent — and traineeships were plentiful. In real terms, funding decreased between 1968 and 1974, increased slightly at the end of the decade, fell in the early 1980s, and then began a modest upward trend in 1983.

Both the "go" and the "stop" parts of the policy have had adverse effects for science. First, more funding spurred phenomenal growth in the numbers of Ph.D.'s. Such growth was accomplished at the expense of quality, as science became less selective. Second, when the "stop" came, new Ph.D.'s had great difficulty getting jobs that fostered research. Instead, many took jobs at second- and third-tier universities, where they have struggled to set up labs, attract students, and do research. Those who could not find jobs in the academic sector went to work for industry or the government, just at the time when resources for research in these sectors became tighter. Careers in science became less appealing. Growth slowed. The scientific community began to age. Just as important, poor job prospects, coupled with a decline in traineeships and fellowships, discouraged many of the most capable students of the 1970s and 1980s from choosing careers in science. The tightness of resources made science more competitive. In all these ways a defective national funding policy contributed to making the current scientific community less productive than it otherwise might have been.

Where Should We Go from Here?

In 1991 Leon Lederman, at the time the president-elect of the American Association for the Advancement of Science, wrote a report entitled "Science: The End of the Frontier?"[31] that detailed the malaise in science, particularly academic science, that prevails today. Lederman's policy recommendation was to double the amount of federal funding available to university researchers over the next two to three years and thereafter increase funding at a real annual rate of between 8 and 10 percent.

One lesson of this book is that such a policy would be bad for science and, by inference, bad for the United States. The double-plus-10-percent solution, if achieved, would fuel an immense growth in science. But rapid growth is not necessarily good for science. Although a dramatic increase in funding would in all likelihood initially attract the cream back into science, too-rapid growth would eventually be accompanied by a decline in the quality of students entering science, the result of digging deeper into the talent distribution. Too-rapid growth would also encourage universities to expand, pursuing fundable projects that involve big teams and large institutional overheads, not necessarily the type of research that fosters good science. The other reason that such a policy would be bad for science is that it is extremely unlikely that it could be sustained. Lederman's recommendation implies a fourfold increase in real spending by the year 2001, a sixfold increase by 2005. Such enormous increases are just not in the political or economic cards. Real funding increases would eventually be cut, perhaps drastically. We then would be back where we started, with another boom–bust cycle, with all the adverse long-run consequences of the last one.

ELEMENTS OF A RATIONAL SCIENCE POLICY

A rational science policy must not repeat the mistakes of the past. Instead, a rational science policy must recognize that science is extremely sensitive to the signals that funding sends. A policy of +15 percent this year and −2 percent next year sends shock waves through science that can be felt for years. To nourish scientific productivity, the United States should commit to long-term real growth in federal funding in the neighborhood of 4 to 5 percent, rather than 8 to 10 percent. Such a policy is much more likely to be sustained and would not generate excessive growth. By giving the scientific community some long-term security, it would also encourage it to take a longer view than has otherwise been possible.

There are other things that can be done to improve the quality of science in the United States. Some of these relate to seeking alternative or supplementary sources of funding. Others relate to reorganizing resources already available in order to use them more efficiently. In the final pages of this book, we outline some of these changes.

Who Should Pay?

Given the nature of basic science, it is unwise to take a strictly national viewpoint. More basic science for country X means more basic science for country Y (and Z). Therefore, basic science should be broadly supported by the world economies that can afford it and stand to benefit from it. This clearly includes Europe and Japan as well as the developing Pacific Rim countries. As a country, we cannot continue to give away basic science, thereby subsidizing world growth. National pride may have to take second place to economic reality. The commitment of the administration to raise 18 percent of the cost of the Superconducting Supercollider from foreign governments is a step in the right direction. Perhaps a country that learned to pass the hat to finance the war in the Persian Gulf can bring itself to do the same for basic science, which benefits all.

More Applied Research

Concern with being first to reach the basic science frontier has led the United States to be slow to collect the rewards associated with basic science breakthroughs. It is time that the United States cash in on more of its basic discoveries. "The challenge for [U.S.] science is learning how to translate scientific breakthroughs into technological breakthroughs. We have the premier scientific community in the world. But that community must believe that it's important and prestigious to make those translations."[32] For many years, the Japanese have fished with great success in the pool of basic science filled by others. They have quickly transformed discoveries into high-quality products and processes for designing, manufacturing, and distributing such products.

To promote applied research, partnerships between research institutions and business must be encouraged. Because business and universities can appropriate many of the benefits of applied science, funding for academic research need not come entirely from governmental sources. Entrepreneurial activities shared by universities, businesses, and professors can be successful. Witness what is happening in biotechnology and pharmacology in which in exchange for certain property rights, industry has helped fund university labs.[33] The promotion of applied research also requires that scientists working in applied areas be awarded prestige by their peers. Careers in industry should not be viewed as second best to those in academe. Progress is being made in this respect, especially in biotechnology, in which careers in the private sector are no longer viewed as distinctly inferior. On the other hand, if industry is to attract top-flight scientists and use their knowledge to translate from basic to applied science, it must allow scientists the freedom to explore creative research agendas. Industry must take a longer view, funding projects that have the potential of a significant payoff, but not necessarily in the next two or three years. The myopic nature of America's R & D enterprise needs to be adjusted. Other countries, notably Japan, have benefited from taking a longer view.

What Should Be Funded?

Funding of individual open-ended research projects should not be shortchanged in order to fund targeted missions. There is no guarantee that targeted dollars create results, and the bureaucracy and large teams that often accompany targeted, managed projects can take the fun out of doing science. Discovery cannot be mandated or controlled. The best we can hope to do is to develop an environment that nurtures a passion for science.

Funding only massive projects is also not a rational strategy. Big Science has become important partly because it is an effective way to harness the national interest and hence gain support for funding. The golden age of science was touched off in large part by a humiliated nation that could not stand the implication that it was second to the Soviet Union. There is a place for Big Science, but much remains to be learned through tabletop science and investigator-initiated "mom and pop" awards.[34] What one physicist called GULPs (grandiose, unnecessary, large projects) should not be allowed to cannibalize SIPs (single-investigator projects).[35] Indeed, "Congress may well consider it better to pay for NASA [America's space program] than for basic research in universities but it should not kid itself that this is the cleverest way to spend money in science."[36] Paul Chu's breakthrough in the field of high-temperature superconductivity in 1987 came from a laboratory operating on less than $150,000.[37]

Who Should Be Funded?

Funding everyone may be politically attractive, but it is not efficient. Science is inherently elite in the sense that few have what it takes to do inspired work, and the "golden" apples ready to be picked at any given time are limited. Even if we as a country could afford to have two hundred Harvards, it is not at all clear that this would make economic sense. More money does not always translate into more results, at least in terms of what could be considered important results.[38] The reward structure that has evolved in science encourages too many institutions and persons to seek research funding and to be disappointed if they do not get it.

Reorganization at Universities

The United States must rethink the way in which university-based research is organized and conducted. Research requires assistants, and this means, for the most part, doctoral students and postdoctorates, whose labor is cheap. The system encourages new Ph.D. programs to be established where there may not be a need and expanded when they already exist, even if graduates are having a difficult time finding research jobs. Without a continual flow of graduate students, the research system as we know it at U.S. universities would not exist. Yet the situation cannot continue forever in which a professor trains three or four students, at least half of whom upon completing their degrees seek positions in academe where they in turn will each recruit three or four more students to work in their labs. Neither the demography nor the resources of this country can support such an agenda. We already have a situation in which more and more of the new Ph.D.'s end up in "unfaculty" positions, positions in which they work as researchers at a university, but without the possibility of getting tenure or, in most instances, the opportunity to be principal investigators on funded research projects.

Perhaps more importantly, it is not clear that doctoral education as it is practiced in the United States today encourages creativity and quality. In many large labs, graduate students are simply assigned a project and told how to go about doing it. The labs become factories. The input of graduate students is "solely in the form of a critical analysis of the methodologies, not in the high-level analysis of the data and the casting of hypotheses, which are the exclusive domain of the laboratory director."[39] Such a system leads scientists to be trained as workers, not thinkers. A fundamental reform would include putting an end to graduate student "slave labor" and reducing the number of Ph.D.'s granted. The master's degree would be revitalized and given to persons who would become permanent research assistants; the Ph.D. would be reserved for the truly talented and creative members of society.

A downsizing of Ph.D. programs would also enable universities to allocate more resources to undergraduate education, in which grave problems exist. A recent study, for example, found that the way in which introductory science courses are taught in college repels "many otherwise intelligent, curious, and ambitious young people."[40] As causes of the defections, the study blames poor teaching, including an overemphasis on the "how" and a neglect of the "why" in science, as well as the tense competitive atmosphere in the science classroom. The reward system in colleges and universities has also stymied quality teaching. The importance of external funding has led universities to reward fundable persons and to punish those who do not get funding, giving the latter heftier teaching loads and smaller raises. In the process the message has become clear: Teaching is to be relegated to those who have no science left in them or to graduate students struggling to begin their careers.[41]

Universities should also reconsider the way in which they allocate resources. Quality research should be rewarded over quantity. Universities should focus on areas of excellence rather than trying to be excellent across the board. Just as it is not cost effective for all hospitals to provide across-the-board healthcare facilities, it has become economically inefficient for most universities to strive for excellence in all areas.[42] Institutions of higher education must also change their expectations of faculty.[43] Requiring faculty in the sciences to have written more than fifty articles in order to get tenure and to have a continual history of funding is counterproductive. It encourages fraud and mediocrity and discourages excellence in teaching. Furthermore, institutions must recognize that federal government funding cannot provide institutional salvation. The funding situation in the late 1950s and 1960s was an anomaly, not a normal state of affairs. It is economically impossible to "exponentiate forever."

Reform at the Elementary and Secondary Level

Science and math education at the primary and secondary levels in the United States is in a dismal state. U.S. children consistently lag behind children from other developed nations in their ability to do math and science. In a recent study of 13-year-olds in South Korea, the United Kingdom, Spain, Ireland, the United States, and several Canadian provinces, the U.S. children scored the lowest in math and below the mean in science.[44] One consequence of poor science education is that some of the best talent, including a disproportionate number of women and minorities, is lost to science, turned off by a system that educates poorly.

The United States must rethink the way math and science are taught at the primary and secondary levels. Much of this rethinking involves a willingness to pay the salaries necessary to attract good teachers to the classroom. Alternative certification programs can also bring talented individuals back to the classroom. It also involves restructuring science education, with an emphasis on communicating to students the "grandeur and elegance of scientific concepts" and "general principles rather than the esoteric details."[45]

Poor science education also means that Americans are scientifically illiterate, even the highly educated. To wit: At the Harvard commencement exercises in 1987, only two of the twenty-three students queried could explain why it is hotter in summer than in winter.[46]

Creative Options for Older Scientists

In 1986 Congress abolished mandatory retirement for most professions. A category specifically excluded was university professors, who are required to retire by age 70. Beginning in 1994, however, professors will no longer be exempt from the law (unless Congress chooses otherwise), opening up the possibility that faculty will stay at colleges and universities until they die. Such tenacity on the part of older faculty could depress scientific productivity and diminish the options available to younger scientists.

Although the proportion of older faculty deciding to continue their careers may not be large (many faculty choose to retire long before the current mandatory age), this change in policy calls for a reassessment of the options open in the United States to older scientists. The National Research Council task force that studied the retirement issue at the request of Congress recommended that universities and colleges develop special incentives to encourage early retirement. Other possibilities include increased portability of pension and health insurance, permitting older faculty to keep their fringe benefit packages while working in a different environment or sector of the economy. With such portability, for example, older scientists might willingly choose to spend several years teaching at the elementary and secondary levels. On a grander scale, Congress could consider establishing a program analogous to the Peace Corps, offering senior scientists the opportunity to make vital contributions to the nation and to developing countries.[47]

A CONJUNCTION OF NEED AND OPPORTUNITY

The decision to alter the way things are done in science, particularly at institutions of higher education, can be painful. But a window of opportunity exists to smooth the process of change. An older scientific community means that in the near future a disproportionate number of scientists will retire, especially if institutions provide retirement incentives and/or the government encourages older scientists to seek jobs in nontraditional areas. At the same time, the baby-bust generation coming to maturity will provide fewer traditional Ph.D. candidates. William Bowen and Julie Sosa predict that the demand for faculty will exceed the supply in the mathematical and physical sciences by substantial amounts for at least a decade beginning in 1997.[48] Although others question

Conclusion

whether there will be such a large shortage, it is likely that opportunities will be better in the near future for prospective candidates than they were in the 1970s and 1980s.[49] In this environment, institutions, with minimal pain, can reorganize and perhaps phase out certain research units. The prospects of a seller's market will also provide an opportunity to lower the bar on the tenure hurdle, for there should be fewer postdocs waiting in the wings for permanent positions.

Reduced pressure to get grants and more emphasis on teaching, coupled with projected faculty shortages or at least a tighter labor market, could lessen the competition among scientists to produce. The incentives for fraud and paper inflation could diminish. More of the fun might be put back into science. Not only fun, but money. A seller's market should lead to higher salaries in science. The most capable students may once again be attracted into careers in science and engineering.

We believe such changes to be a prescription for better science. We are, by training, economists, and economists are firmly committed to asking whether resources can be more effectively used. We believe that many of the strategies we have described would, through a reallocation of resources, make the scientific enterprise more efficient and hence more productive. But as economists, we also know about the role of choice. The time has long since come for the United States to ask, "Just how much science do we want? How many centers of excellence can we afford?" There are trade-offs. More dollars for science mean fewer dollars for other, pressing social needs. But it is not as simple as that. More dollars for science also mean economic growth and thus more dollars for other things in the future. Science is an investment that expands the opportunities open to a country and the world. The effects of sporadic policies, as we have seen, can be felt for decades. The choices to be made with regard to science policy are complex. To make these kinds of decisions, however, or at least to make them well, requires informed, educated policymakers and an American electorate that appreciates the importance of science. In the decades ahead, America must become scientifically literate.

NOTES

1. Special SDR tabulations for U.S.-trained doctoral scientists employed full time in science or engineering in 1973 and 1987. The sciences include all the physical and life sciences but exclude engineering, psychology, and the social sciences. Academe includes four-year colleges, universities, and medical schools.
2. Planck 1949, pp. 33–34.
3. Interviewed on the "MacNeil/Lehrer News Hour," October 2, 1987.
4. William Booth, "Science and the Art of Money," *Washington Post*, February 17, 1991.
5. Jeffrey Mervis, "Panel Weighs Overhaul of NSF's Grant System," *The Scientist*, January 7, 1991.
6. Dana Milbank, "Research Setback," *Wall Street Journal*, November 7, 1990.
7. Szilagyi and Zell 1990, p. 92.
8. Not only is there more pressure on scientists to seek funds, but there are also more scientists than ever before as potential applicants. Although science has not grown in recent years at the rate it did in the 1960s, it has nonetheless continued to grow. In 1987, for example, there were approxi-

mately 175,000 more doctoral scientists (including social scientists but excluding engineers) than there were in 1960 (see National Research Council 1978, 1988).

9. The length of the downturn in the early 1970s depends on the price deflator used and how federal expenditures are counted (see Lederman 1991, p. 8; U.S. Congress 1991, p. 6).

10. U.S. Congress 1991, p. 22. Reimbursement for indirect costs is the fastest-growing portion of federal research expenditures. Recently there has been a good deal of concern over whether all costs attributed to overhead are legitimate (see Palca 1991).

11. U.S. Congress 1991, p. 23.

12. Lederman 1991.

13. Some would argue that the greater emphasis on regulation, which requires researchers to comply with certain guidelines, is another reason that the cost of doing science has risen.

14. Hackett 1990, p. 264.

15. Broad 1981.

16. Hackett 1990, p. 263.

17. Ibid.

18. Ibid.

19. Lederman 1991.

20. For additional discussion concerning the issue of risk, see Jeffrey Mervis, "Panel Weighs Overhaul of NSF's Grant System," *The Scientist*, January 7, 1991.

21. Ibid.

22. For example, one scientist reported that in a grant application he suppressed a discovery (even though it was soon to be published in a prestigious journal) because he feared the granting agency would not believe the breakthrough results and hence would give his proposal a poor priority score (see Hackett 1990, p. 264).

23. See Kohn 1988, pp. 52-57, 131-41. The Piltdown man hoax centered on the discovery of fragments of a humanlike brain and an apish jaw that were alleged to be proof of the missing link in the evolutionary chain between ape and human.

24. Chubin 1988.

25. For a discussion of this incident, see, for example, Culliton 1989, pp. 643-6, and Philip J. Hilts, "How Charges of Lab Fraud Grew Into a Cause Célèbre," *New York Times*, March 26, 1991.

26. This issue drew national attention when Walter Stewart and Ned Feder carefully examined 109 papers in clinical and experimental cardiology coauthored by John Darsee, "looking for published errors, seemingly misleading statements, and other defects" (see Chubin 1988, p. 57). Stewart and Feder's (1987) study took several years to appear in print because of the threats by the lawyers representing the coauthors.

27. Close 1991.

28. Feynman 1985, p. 158.

29. Adams 1990b, p. 693, found that about two thirds of average (between 1953 and 1980) multifactor productivity growth is accounted for by increases in scientific knowledge occurring twenty years earlier.

30. There is some evidence that compared with earlier periods, it now takes a shorter time to reap the economic benefits of basic science. This appears to be particularly true in molecular biology. For example, within six years in the 1970s, fundamental experiments with recombinant-DNA techniques were translated into genetically engineered insulin.

31. Lederman 1991.

32. Mary Good, chemist and chairwoman of the National Science Board, quoted in Julia King, "Scientists Wary as New Year Dawns," *The Scientist*, January 7, 1991, p. 14.

33. See Etzkowitz 1983.

34. See Dalrymple 1991 for a discussion of the importance of "small" science.

35. Kadin 1990.

36. "Money for the Boffins," *The Economist*, February 16, 1991, p. 16.

37. William J. Broad, "Vast Sums for New Discoveries Pose a Threat to Basic Science," *New York Times*, May 27, 1990.

38. There is a significant amount of evidence that only a small fraction of scientific work and, by inference, of those doing the work, is outstanding. For example, it is common to state that

Conclusion 173

about 10 percent of the proposals submitted for funding are of outstanding quality (see Abelson 1991). And for many years it has been recognized that only about 6 percent of publishing scientists write 50 percent of all papers (see Stephan and Levin 1991).

39. Letter from Professor Robert E. Hurst, College of Medicine, University of Oklahoma.

40. Tobias 1990, p. 11.

41. In some instances, large lecture sections taught by eminent scientists do exist, but hands-on-instruction in the labs is generally given by graduate teaching assistants. There is already some indication that higher education is beginning to reevaluate the way in which resources are divided between research and teaching. See Karen Grassmuck, "Some Research Universities Contemplate Sweeping Changes, Ranging from Management and Tenure to Teaching Methods," *Chronicle of Higher Education*, September 12, 1990.

42. A recent study by the Office of Technology Assessment made a similar point when it recommended that "have-not" institutions concentrate on select areas of research rather than attempting to be excellent across the board (see U.S. Congress 1991, p. 31).

43. An example of such a change recently occurred at Harvard, where the medical school has decided to limit to ten the number of articles that the promotion and tenure committee will review.

44. National Science Board 1989, p. 7.

45. See, for example, John S. Rigden and Sheila Tobias, "Too Often, College-Level Science Is Dull as Well as Difficult," *Chronicle of Higher Education*, March 27, 1991; and Robert M. Hazen and James Trefil, "General Science Courses Are the Key to Scientific Literacy," *Chronicle of Higher Education*, April 10, 1991.

46. Hazen and Trefil 1991, p. xiv. It is not only the highly educated that are scientifically illiterate. Working scientists are often illiterate outside their own fields of professional expertise. Hazen and Trefil (1991, p. xiii) report that out of twenty-four physicists and geologists asked to explain the difference between DNA and RNA, only three could do so, and all three were doing research in an area in which knowledge of the difference would be useful.

47. We owe this suggestion to John S. Rigden (personal communication).

48. Bowen and Sosa (1989, p. 218) provide projections of faculty shortages using four different scenarios. For the biological sciences and psychology combined, they found that the market will noticeably tighten but still remain in a state of excess supply throughout the period 1987 to 2007. Data for the biological sciences alone are not presented. Others agree that a shortage will occur (see Vaughn and Rosenzweig 1990).

49. Jeffrey Mervis, "Analysts Debunk Idea of Scientist Shortage, Citing Defects in Current Economic Models," *The Scientist*, April 29, 1991. Also see Jeffrey Mervis, "Pundits Foresee Stiffer Job Competition in Academia," *The Scientist*, May 13, 1991.

References

Abelson, Philip H. 1991. "Research Funding." *Science* 252:625.
Adams, James D. 1990a. "Efficient Funding of Scientific Research: An Experiment in Applied Welfare Economics." Paper presented to the meeting of the American Economic Association, Washington, DC.
———. 1990b. "Fundamental Stocks of Knowledge and Productivity Growth." *Journal of Political Economy* 98:673–702.
Ahlburg, Dennis, Eileen M. Crimmins, and Richard A. Easterlin. 1981. "The Outlook for Higher Education: A Cohort Size Model of Enrollment of the College Age Population, 1948–2000." *Review of Public Data Use* 9:211–27.
Albert, Robert S. 1990. "Identity, Experiences and Career Choice Among the Exceptionally Gifted and Eminent." In *Theories of Creativity*, ed. Mark A. Runco and Robert S. Albert, pp. 13–34. Newbury Park, CA: Sage.
Allison, Paul D., J. Scott Long, and Tad K. Krauze. 1982. "Cumulative Advantage and Inequality in Science." *American Sociological Review* 47:615–25.
Allison, Paul D., and John A. Stewart. 1974. "Productivity Differences Among Scientists: Evidence for Accumulative Advantage." *American Sociological Review* 39:596–606.
Alpher, R. A., M. D. Fiske, F. S. Ham, and P. B. Kahn. 1979. "Summary of a Statistical Study of the Ph.D. Physicist Employed in Industry." In *The Transition in Physics Doctoral Employment 1960–1990*, pp. 25–32. Report of the Physics Manpower Panel of the American Physical Society. New York: American Physical Society.
Amabile, Teresa M. 1990. "Within You, Without You: The Social Psychology of Creativity, and Beyond." In *Theories of Creativity*, ed. Mark A. Runco and Robert S. Albert, pp. 61–91. Newbury, CA: Sage.
Amemiya, Takeshi. 1981. "Qualitative Response Models: A Survey." *Journal of Economic Literature* 19:483–536.
Andrews, Frank M. ed. 1979. *Scientific Productivity, the Effectiveness of Research Groups in Six Countries*. Cambridge: Cambridge University Press.
Astin, Alexander. 1971. *Academic Performance in College*. New York: Free Press.
Austin, James H. 1978. *Chase, Chance and Creativity: The Lucky Art of Novelty*. New York: Columbia University Press.
Barash, David P. 1983. *Aging: An Exploration*. Seattle: University of Washington Press.
Barber, Bernard. 1962. "Resistance by Scientists to Scientific Discovery." In *The Sociology of Science*, ed. Bernard Barber and Walter Hirsch, pp. 539–56. New York: Free Press.
Barber, Bernard, and Renee C. Fox. 1962. "The Case of the Floppy-eared Rabbits: An Instance of Serendipity Gained and Serendipity Lost." In *The Sociology of Science*, ed. Bernard Barber and Walter Hirsch, pp. 525–38. New York: Free Press.
Barron, Frank, and David M. Harrington. 1981. "Creativity, Intelligence, and Personality." *Annual Reviews of Psychology* 32:439–76.
Bayer, Alan E., and Jeffrey C. Dutton. 1977. "Career Age and Research-Professional Activities of Academic Scientists." *Journal of Higher Education* 48:259–82.

Bayer, Alan E., and John K. Folger. 1966. "Some Correlates of Citation Measure of Productivity in Science." *Sociology of Education* 39:381-90.
Beard, George M. 1874. *Legal Responsibility in Old Age*. New York: Russell.
Becker, Gary S. 1964. *Human Capital*. New York: Columbia University Press for the National Bureau of Economic Research.
Ben-Porath, Yoram. 1967. "The Production of Human Capital and the Life Cycle of Earnings." *Journal of Political Economy* 75:352-65.
Birren, James E., Anita M. Woods, and M. Virtrue Williams. 1980. "Behavioral Slowing with Age: Causes, Organization, and Consequences." In *Aging in the 1980s: Psychological Issues*, ed. Leonard W. Poon, pp. 293-308. Washington, DC: American Psychological Association.
Bishop, John H. 1989. "Is the Test Score Decline Responsible for the Productivity Growth Decline?" *American Economic Review* 79:178-97.
Blackburn, Robert T., Charles E. Behymer, and D. E. Hall, 1978. "Correlates of Faculty Publications." *Sociology of Education* 51:132-41.
Blau, Judith R. 1978. "Sociometric Structure of a Scientific Discipline." In *Research in Sociology of Knowledge, Sciences and Art*, ed. Robert A. Jones, pp. 191-206. Greenwich, CT: JAI Press.
Blau, Peter. 1973. *The Organization of Academic Work*. New York: Wiley.
Bliss, Michael. 1982. *The Discovery of Insulin*. Chicago: University of Chicago Press.
Bowen, Howard R., and Jack H. Schuster. 1986. *American Professors: A National Resource Imperiled*. New York: Oxford University Press.
Bowen, William G., and Julie Ann Sosa. 1989. *A Study of Factors Affecting Demand and Supply, 1987 to 2012*. Princeton, NJ: Princeton University Press.
Broad, William J. 1981. "The Publishing Game: Getting More for Less." *Science* 211:1137-9.
Brodetsky, S. 1942. "Newton: Scientist and Man." *Nature* 150:698-9.
Busse, Ewald W. 1989. "The Myth, History, and Science of Aging." In *Geriatric Psychiatry*, ed. Ewald W. Busse and Dan G. Blazer, pp. 3-34. Washington DC: American Psychiatric Press.
Caplow, Theodore, and Reece J. McGee. 1958. *The Academic Marketplace*. New York: Basic Books.
Carlsson, Gosta, and Katarina Karlsson. 1970. "Age Cohorts and Generation of Generations." *American Sociological Review* 35:710-18.
Carnegie Foundation for the Advancement of Teaching. 1987. *A Classification of Institutions of Higher Education*. Princeton, NJ: Carnegie Foundation Technical Report.
Cartter, Allan M. 1966. *An Assessment of Quality in Graduate Education*. Washington, DC: American Council on Education.
Cattell, Raymond B. 1963. "The Personality and Motivation of the Researcher from Measurements of Contemporaries and from Biography." In *Scientific Creativity: Its Recognition and Development*, ed. Calvin W. Taylor and Frank Barron, pp. 119-32. New York: Wiley.
Chubin, Daryl E. 1988. "Allocating Credit and Blame in Science." *Science, Technology & Human Values* 13:53-63.
Cohen, I. Bernard. 1985. *Revolution in Science*. Cambridge, MA: Harvard University Press.
Close, Frank. 1991. *Too Hot to Handle*. Princeton, NJ: Princeton University Press.
Cole, Gerald A. 1979. "Classifying Research Units by Patterns of Performance and Influence: A Typology of the Round I Data." In *Scientific Productivity, The Effectiveness of Research Groups in Six Countries*, ed. Frank M. Andrews, pp. 353-404. Cambridge: Cambridge University Press.
Cole, Jonathan R., and Stephen Cole. 1973. *Social Stratification in Science*. Chicago: University of Chicago Press.
Cole, Stephen. 1979. "Age and Scientific Performance." *American Journal of Sociology* 84:958-77.
Cole, Stephen, and G. S. Meyer. 1985. "Little Science, Big Science Revisited." *Scientometrics* 7:443-58.
Crane, Diana. 1972. *Invisible Colleges: Diffusion of Knowledge in Scientific Communities*. Chicago: University of Chicago Press.
Crick, Francis. 1988. *What Mad Pursuit: A Personal View of Scientific Discovery*. New York: Basic Books.
Cullison, William E. 1989. "The U.S. Productivity Slowdown: What the Experts Say." *Economic Review* (Federal Reserve Bank of Richmond), July-August, pp. 10-21.
Culliton, Barbara J. 1989. "The Dingell Probe Finally Goes Public." *Science* 244:643-6.
Dalrymple, G. Brent. 1991. "The Importance of 'Small' Science." *Eos* 72:1,4.

References

Dannefer, Dale. 1984. "Adult Development and Social Theory: A Paradigmatic Reappraisal." *American Sociological Review* 49:100-16.
Darwin, Charles R. 1859/1966. *On the Origin of Species* (facsimile). Cambridge, MA: Harvard University Press.
Dennis, Wayne. 1954. Review of *Age and Achievement*, by Harvey C. Lehman. *Psychological Bulletin* 5:306-8.
———. 1956a. "Age and Achievement: A Critique." *Journal of Gerontology* 11:331-3.
———. 1956b. "Age and Productivity Among Scientists." *Science* 123:724-5.
———. 1958. "The Age Decrement in Outstanding Scientific Contributions: Fact or Artifact?" *American Psychologist* 13:457-640.
———. 1966. "Creative Productivity Between the Ages of 20 and 80 Years." *Journal of Gerontology* 21:1-8.
Diamond, Arthur M., Jr. 1980. "Age and the Acceptance of Cliometrics." *Journal of Economic History* 40:838-41.
———. 1984. "An Economic Model of the Life-Cycle Research Productivity of Scientists." *Scientometrics* 6:189-96.
———. 1986a. "The Life-Cycle Research Productivity of Mathematicians and Scientists." *Journal of Gerontology* 41:520-5.
———. 1986b. "What Is a Citation Worth?" *Journal of Human Resources* 21:200-15.
Doctorate Records File (DRF). See Survey of Earned Doctorates.
Easterlin, Richard A. 1987. "Easterlin Hypothesis." In *The New Palgrave, a Dictionary of Economics*, ed. John Fatwell, Murray Milgate, and Peter Newman, pp. 1-4. New York: Stockton Press.
The Economist. 1991a. "A Survey of Science." *The Economist*, February 16, special supplement.
———. 1991b. "Money for the Boffins." *The Economist*, February 16, pp. 15-16.
———. 1991c. "Veni, Vidi, Vici." *The Economist*, March 2, pp. 25-26.
Ehrenberg, Ronald G., Charles Clotfelter, Malcom Getz, and John Siegfried. 1991. *Economic Challenges in Higher Education*. Chicago: University of Chicago Press.
Einstein, Albert. n.d. Unpublished manuscript for *Nature*. New York: Morgan Library.
Einstein, Albert, and Leopold Infeld. 1938. *The Evolution of Physics: The Growth of Ideas from Early Concepts to Relativity and Quanta*. New York: Simon & Schuster.
Etzkowitz, Henry. 1983. "Entrepreneurial Scientists and Entrepreneurial Universities in American Academic Science." *Minerva* 21:198-233.
Evenson, Robert E., and Yoav Kislev. 1975. *Agricultural Research and Productivity*. New Haven, CT: Yale University Press.
Feynman, Richard P. 1985. *"Surely You're Joking, Mr. Feynman!"* New York: Bantam.
Fox, Mary F. 1983. "Publication Productivity Among Scientists: A Critical Review." *Social Studies of Science* 13:285-305.
Franklin, Jon. 1987. *Molecules of the Mind*. New York: Dell.
Freeman, Richard B. 1975. "Supply and Salary Adjustments to the Changing Science Manpower Market: Physics, 1948-1973." *American Economic Review* 65:27-39.
Fulton, Oliver, and Martin Trow. 1974. "Research Activity in American Higher Education." *Sociology of Education* 47:29-73.
Gamow, George. 1966. *Thirty Years That Shook Physics: The Story of Quantum Theory*. Garden City, NY: Doubleday.
Garfield, Eugene. 1983. *Citation Indexing, Its Theory and Application in Science, Technology, and Humanities*. Philadelphia: ISI Press.
Garvey, William D., Nan Lin, and Carnot E. Nelson. 1970. "Some Comparisons of Communication Activities in the Physical and Social Sciences." In *Communications Among Scientists and Engineers*, ed. Carnot E. Nelson and Donald K. Pollock, pp. 61-84. Lexington, MA: Heath-Lexington Books.
Gaston, Jerry. 1971. "Secretiveness and Competition for Priority of Discovery in Physics." *Minerva* 9:472-92.
Giere, Ronald N. 1988. *Explaining Science: A Cognitive Approach*. Chicago: University of Chicago Press.
Gieryn, Thomas F., and Richard F. Hirsh. 1983. "Marginality and Innovation in Science." *Social Studies of Science* 13:87-106.
Gillmor, C. Stewart. 1984. "Aging of Geophysicists." *Eos* 65:353-4.
———. 1987. "Aging of Geophysicists Reconsidered." *Eos* 68:802-5.
Gleick, James. 1987. *Chaos: Making a New Science*. New York: Viking.

Glen, William. 1982. *The Road to Jaramillo*. Stanford, CA: Stanford University Press.
Gottlieb, Annie. 1987. *Do You Believe in Magic? (The Second Coming of the Sixties Generation)*. New York: New York Times Books.
Gribbin, John R. 1987. *In Search of the Double Helix*. New York: Bantam.
Hackett, Edward J. 1990. "Science as a Vocation in the 1990's: The Changing Organizational Culture of Academic Science." *Journal of Higher Education* 61:233–79.
Hagstrom, Warren O. 1965. *The Scientific Community*. New York: Basic Books.
Hargens, Lowell L. 1975. *Patterns of Scientific Research: A Comparative Analysis of Research in Three Scientific Fields*. Washington, DC: American Sociological Association.
———. 1988. "Scholarly Consensus and Journal Rejection Rates." *American Sociological Review* 53:139–51.
Harmon, Lindsey R. 1961. "The High School Backgrounds of Science Doctorates." *Science* 133:679–88.
Harré, Rom. 1979. *Social Being*. Oxford: Basil Blackwell.
Harrington, David M. 1990. "The Ecology of Human Creativity: A Psychological Perspective." In *Theories of Creativity*, ed. Mark A. Runco and Robert S. Albert, pp. 143–69. Newbury Park, CA: Sage.
Hartnett, Rodney T. 1987. "Has There Been a Graduate Student 'Brain Drain' in the Arts and Sciences?" *Journal of Higher Education* 58:562–85.
Hausman, Jerry A., and William E. Taylor. 1981. "Panel Data and Unobserved Individual Effects." *Econometrica* 49:1377–98.
Hawking, Stephen W. 1990. *A Brief History of Time*. New York: Bantam.
Hazen, Robert M. 1988. *The Breakthrough: The Race for the Superconductor*. New York: Ballantine.
Hazen, Robert M., and James Trefil. 1991. *Science Matters*. Garden City, NY: Doubleday.
Heckman, James J. 1976. "The Common Structure of Statistical Models of Truncation, Sample Selection and Limited Dependent Variables and a Simple Estimator for Such Models." *Annals of Economic and Social Measurement* 5:575–9.
Heckman, James J., and Thomas E. MaCurdy. 1980. "A Life Cycle Model of Female Labor Supply." *Review of Economic Studies* 47:47–74.
Hull, David L. 1988. *Science as a Process*. Chicago: University of Chicago Press.
Hull, David L., Peter D. Tessner, and Arthur M. Diamond. 1978. "Planck's Principle." *Science* 202:717–23.
Institute for Scientific Information. Annual. *Science Citation Index (SCI)*. Philadelphia: Institute for Scientific Information.
Institute for Scientific Information. Various dates. *SCI Journal Citation Reports*. Philadelphia: Institute for Scientific Information.
Jones, Lyle V., Gardner Lindzey, and Porter Coggeshall, eds. 1982. *An Assessment of Research Doctorate Programs in the U.S.: Mathematical and Physical Sciences*. Washington, DC: National Academy Press.
Judge, George G., William E. Griffiths, R. Carter Hill, and Tsoung-Chao Lee. 1980. *The Theory and Practice of Econometrics*. New York: Wiley.
Kadin, Alan M. 1990. Letter in *Physics Today* 43 (December): 92.
Kanigel, Robert. 1986. *Apprentice to Genius: The Making of a Scientific Discovery*. New York: Macmillan.
Keniston, Hayward. 1959. *Graduate Study and Research in the Arts and Sciences at the University of Pennsylvania*. Philadelphia: University of Pennsylvania Press.
Killingsworth, Mark R. 1983. *Labor Supply*. Cambridge: Cambridge University Press.
Koch, H. William. 1971. "On Physics and Employment of Physicists in 1970." *Physics Today* 24(June): 23–27.
Kohn, Alexander. 1988. *False Prophets*. New York: Basil Blackwell.
Kuhn, Thomas S. 1970. *The Structure of Scientific Revolutions*. 2d ed. Chicago: University of Chicago Press.
Kyvik, Svein. 1990. "Age and Scientific Productivity: Differences Between Fields of Learning." *Higher Education* 19:37–55.
Law, John. 1980. "Fragmentation and Investment in Sedimentology." *Social Studies of Science* 10:1–22.
Lawani, S. M. 1986. "Some Bibliometric Correlates of Quality in Scientific Research." *Scientometrics* 9:13–25.
Lawrence, Janet H., and Robert T. Blackburn. 1988. "Age as a Predictor of Faculty Productivity: Three Conceptual Approaches." *Journal of Higher Education* 59:22–38.

Lederman, Leon M. 1991. "Science: The End of the Frontier?" *Science*, January, supplement.
Lehman, Harvey C. 1944. "Man's Most Creative Years: Quality Versus Quantity of Output." *Scientific Monthly* 59:384-93.
———. 1953. *Age and Achievement*. Princeton, NJ: Princeton University Press.
———. 1956. "Reply to Dennis' Critique of Age and Achievement." *Journal of Gerontology* 11:333-7.
———. 1958. "The Chemist's Most Creative Years." *Science* 127:1213-21.
———. 1960. "The Age Decrement in Outstanding Scientific Creativity." *The American Psychologist* 15:128-34.
———. 1962a. "The Creative Production Rates of Present Versus Past Generations of Scientists." *Scientific Productivity* 17:409-17.
———. 1962b. "More About Age and Achievement." *Gerontologist* 2:141-8.
———. 1963. "Chronological Age Versus Present-Day Contributions to Medical Progress." *Gerontologist* 3:71-75.
Levi-Montalcini, Rita. 1988. *In Praise of Imperfection: My Life and Work*. New York: Basic Books.
Levin, Sharon G., and Paula E. Stephan. 1991. "Research Productivity over the Life Cycle: Evidence for Academic Scientists." *American Economic Review* 81:114-32.
Levinson, Daniel J. 1977. "The Mid-Life Transition: A Period in Adult Psychosocial Development." *Psychiatry* 40:99-112.
———. 1978. *The Seasons of a Man's Life*. New York: Knopf.
Lindsey, Duncan. 1980. "Production and Citation Measures in the Sociology of Science: The Problem of Multiple Authorship." *Social Studies of Science* 10:145-62.
Lodahl, Janice B., and Gerald Gordon. 1972. "The Structure of Scientific Fields and the Functioning of University Graduate Departments." *American Sociological Review* 37:57-72.
Long, J. Scott. 1978. "Productivity and Academic Position in the Scientific Career." *American Sociological Review* 43:889-908.
Long, J. Scott, Paul D. Allison, and Robert McGinnis. 1979. "Entrance into the Academic Career." *American Sociological Review* 44:816-30.
Long, J. Scott, and Robert McGinnis. 1981. "Organizational Context and Scientific Productivity." *American Sociological Review* 46:422-42.
Maddala, G. S. 1983. *Limited-Dependent and Qualitative Variables in Econometrics*. Cambridge: Cambridge University Press.
———. 1987. "Limited Dependent Variable Models Using Panel Data." *Journal of Human Resources* 22:307-38.
Manniche, E., and G. Falk. 1957. "Age and the Nobel Prize." *Behavioral Science* 2:301-7.
Mansfield, Edwin. 1991. "Academic Research and Industrial Innovation." *Research Policy* 20:1-12.
McCann, H. Gilman. 1978. *Chemistry Transformed*. Norwood, NJ: Ablex.
McDowell, John M. 1982. "Obsolescence of Knowledge and Career Publication Profiles: Some Evidence of Differences Among Fields in Costs of Interrupted Careers." *American Economic Review* 72:752-68.
McFadden, Daniel. 1974. "Conditional Logit Analysis of Qualitative Choice Behavior." In *Frontiers in Econometrics*, ed. Paul Zarembka, pp. 105-42. New York: Academic Press.
Menard, Henry W. 1971. *Science: Growth and Change*. Cambridge, MA: Harvard University Press.
Merton, Robert K. 1957. "Priorities in Scientific Discovery: A Chapter in the Sociology of Science." *American Sociological Review* 22:635-59.
———. 1961. "Singletons and Multiples in Scientific Discovery." *Proceedings of the American Philosophical Society* 105:470-86.
———. 1968. "The Matthew Effect in Science." *Science* 159:56-63.
Messeri, Peter. 1988. "Age Differences in the Reception of New Scientific Theories: The Case of Plate Tectonics Theory." *Social Studies of Science* 18:91-112.
Miller, A. Carolyn, and Sharon L. Serzan. 1984. "Criteria for Identifying a Refereed Journal." *Journal of Higher Education* 6:673-97.
Mincer, Jacob. 1974. *Schooling, Experience, and Earnings*. New York: Columbia University Press for the National Bureau of Economic Research.
Mitroff, Ian I. 1974. "Norms and Counter-Norms in a Select Group of Apollo Moon Scientists: A Case Study of the Ambivalence of Scientists." *American Sociological Review* 39:579-95.
Mokyr, Joel. 1990. *The Lever of Riches*. New York: Oxford University Press.
Narin, Francis. 1976. *Evaluative Bibliometrics: The Use of Citation Analysis in the Evaluation of Scientific Activity*. Cherry Hill, NJ: Computer Horizons.
National Institute of Mental Health. 1971. *Human Aging I: A Biological and Behavioral Study*.

U.S. Department of Health, Education and Welfare. Washington, DC: U.S. Government Printing Office.
National Merit Scholarship Corporation. Various dates. *Annual Report*.
National Research Council. 1965. *Profiles of PhD's in the Sciences*. Washington, DC: National Academy of Sciences. *Summary Report on Follow-Up of Doctorate Cohorts 1935-1960*.
———. 1972. *Physics in Perspective*. Vol 1. Washington, DC: National Academy of Sciences.
———. 1978. *A Century of Doctorates: Data Analysis of Growth and Change*. Washington, DC: National Academy of Sciences.
———. 1988. *Characteristics of Doctoral Scientists and Engineers in the United States: 1987*. Washington, DC: National Science Foundation (NSF 88-331).
———. Various dates. *Science, Engineering, and Humanities Doctorates in the United States*. Washington, DC: National Academy Press. (Based on the Survey of Doctorate Recipients, SDR.)
———. Various dates. *Doctorate Recipients from United States Universities*. Washington, DC: National Academy Press. (Based on the Survey of Earned Doctorates, SED.)
National Science Board. 1977. *Report to the Subcommittee in Science, Research and Technology, U.S. House of Representatives Regarding Peer Review Procedures at the National Science Foundation*. Washington, DC: U.S. Government Printing Office (NSB 77-468).
———. 1987. *Science & Engineering Indicators—1987*. Washington, DC: U.S. Government Printing Office (NSB 87-1).
———. 1989. *Science & Engineering Indicators—1989*. Washington, DC: U.S. Government Printing Office (NSB 89-1).
National Science Foundation. 1990a. *Great Achievements, Great Expectations*. National Science Foundation 40th Anniversary Symposium: A Report. Washington, DC: National Science Foundation (NSF 90-63).
———. 1990b. *Science and Engineering Doctorates: 1960-89*. Washington, DC: National Science Foundation (NSF 90-320).
Nelson, Helen. 1928. "The Creative Years." *American Journal of Psychology* 40:303-11.
Neugarten, Bernice L. 1968a. "Adult Personality: Towards a Psychology of the Life Cycle." In *Middle Age and Aging: A Reader in Social Psychology*, ed. Bernice L. Neugarten, pp. 137-47. Chicago: University of Chicago Press.
———. 1968b. "The Awareness of Middle Age." In *Middle Age and Aging: A Reader in Social Psychology*, ed. Bernice L. Neugarten, pp. 93-98. Chicago: University of Chicago Press.
Nordhaus, William D. 1982. "Economic Policy in the Face of Declining Productivity Growth." *European Economic Review* 18:131-58.
Nulty, Peter. 1989. "The Hot Demand for New Scientists." *Fortune*, July 31, pp. 155, 158, 162-3.
Olsen, Randall J. 1980. "A Least Squares Correction for Selection Bias." *Econometrica* 48: 1815-20.
Pais, Abraham. 1986. *Inward Bound: Of Matter and Forces in the Physical World*. New York: Oxford University Press.
Palca, Joseph. 1990. "Young Investigators at Risk." *Science* 249:351-3.
———. 1991. "Indirect Costs: The Gathering Storm." *Science* 252:636-8.
Panem, Sandra. 1984. *The Interferon Crusade*. Washington, DC: Brookings Institution.
Pelz, Donald C., and Frank M. Andrews. 1976. *Scientists in Organizations*. Rev. ed. Ann Arbor: Institute for Social Research, University of Michigan.
Pendlebury, David A. 1991. Letter in *Science* 251:1410-11.
Pickering, Andrew. 1984. *Constructing Quarks*. Chicago: University of Chicago Press.
Planck, Max. 1949. *Scientific Autobiography and Other Papers*. New York: Philosophical Library.
Poon, Leonard W. 1985. "Differences in Human Memory with Aging: Nature, Causes, and Clinical Implications." In *Handbook of The Psychology of Aging*, 2d ed., ed. James E. Birren and K. Warner Schaie, pp. 427-62. New York: Van Nostrand Reinhold.
Porter, Beverly Fearn. 1979a. "Mobile Young Faculty; A Follow-Up Study of Untenured Assistant Professors Leaving a Sample of Top Physics Departments." In *The Transition in Physics Doctoral Employment 1960-1990*, pp. 49-112. Report of the Physics Manpower Panel of the American Physical Society. New York: American Physical Society.
———. 1979b. "Transition—A Follow-Up Study of 1973 Postdoctorals." In *The Transition in Physics Doctoral Employment 1960-1990*, pp. 113-92. Report of the Physics Manpower Panel of the American Physical Society. New York: American Physical Society.
Porter, Beverly Fearn, and R. Czujko. 1981. *American Institute of Physics Society Membership, 1981 Profile: An Expanded View*. New York: American Institute of Physics.

Pravdić, Nevenka, and Vesna Oluić-Vuković. 1986. "Dual Approach to Multiple Authorship in the Study of Collaboration/Scientific Output Relationship." *Scientometrics* 10:259-80.
Price, Derek J. de Solla. 1970. "Citation Measures of Hard Science, Soft Science, Technology, and Nonscience." In *Communications Among Scientists and Engineers*, ed. Carnot E. Nelson and Donald K. Pollock, pp. 3-22. Lexington, MA: Heath-Lexington Books.
_____. 1976. "A General Theory of Bibliometric and Other Cumulative Advantage Processes." *Journal of the American Society for Information Science* 27:292-306.
_____. 1986. *Little Science, Big Science . . . And Beyond*. New York: Columbia University Press.
Raup, David M. 1986. *The Nemesis Affair*. New York: Norton.
Reif, Fred, and Anselm Strauss. 1965. "The Impact of Rapid Discovery upon the Scientist's Career." *Social Problems* 12:297-311.
Reskin, Barbara F. 1977. "Scientific Productivity and the Reward Structure of Science." *American Sociological Review* 42:491-504.
_____. 1979. "A Review of the Literature on the Relationship Between Age and Scientific Productivity." In *Research Excellence Through the Year 2000*, pp. 189-207. Washington, DC: National Academy of Sciences.
Rigden, John S. 1987. *Rabi, Scientist and Citizen*. New York: Basic Books.
Roberts, Leslie. 1991. "The Rush to Publish." *Science* 251:260-3.
Roe, Anne. 1952. *The Making of a Scientist*. New York: Dodd Mead.
_____. 1963. "Personal Problems and Science." In *Scientific Creativity: Its Recognition and Development*, ed. Calvin W. Taylor and Frank Barron, pp. 132-8. New York: Wiley.
Roose, Kenneth D., and Charles J. Anderson. 1970. *A Rating of Graduate Programs*. Washington, DC: American Council of Education.
Rosovsky, Henry. 1990. *The University: An Owner's Manual*. New York: Norton.
Runco, Mark A. 1990. "Implicit Theories and Ideational Creativity." In *Theories of Creativity*, ed. Mark A. Runco and Robert S. Albert, pp. 234-52. Newbury Park, CA: Sage.
Salthouse, Timothy A. 1985. "Speed of Behavior and Its Implications for Cognition." In *Handbook of the Psychology of Aging*, 2d ed., ed. James E. Birren and K. Warner Schaie, pp. 400-26. New York: Van Nostrand Reinhold.
Schaie, K. Warner. 1958. "Rigidity-Flexibility and Intelligence: A Cross-sectional Study of the Adult Life Span from 20-70." *Psychological Monographs* 72:1-26.
_____. 1983. "Age Changes in Adult Intelligence." In *Aging: Scientific Perspectives and Social Diseases*, 2d ed., ed. Diana S. Woodruff and James E. Birren, pp. 137-48. Monterey, CA: Brooks/Cole.
Schaie, K. Warner, and Gisela V. Labouvie-Vief. 1974. "Generational Versus Ontogenetic Components of Change in Adult Cognitive Behavior: A Fourteen Year Cross-sequential Study." *Developmental Psychology* 10:305-20.
Scherer, Frederic M. 1986. "The World Productivity Growth Slump." In *Organizing Industrial Development*, ed. Rolf Wolff, pp. 15-27. Berlin: Walter de Gruyter.
Schultz, Theodore W. 1963. *The Economic Value of Education*. New York: Columbia University Press.
Siegler, Ilene C., and Leonard W. Poon. 1989. "The Psychology of Aging." In *Geriatric Psychiatry*, ed. Ewald W. Busse and Dan G. Blazer, pp. 163-201. Washington, DC: American Psychiatric Press.
Simonton, Dean K. 1975. "Sociocultural Context of Individual Creativity: A Transhistorical Time-Series Analysis." *Journal of Personality and Social Psychology* 32:1119-33.
_____. 1976. "Does Sorokin's Data Support His Theory?: A Study of Generational Fluctuations in Philosophical Beliefs." *Journal for the Scientific Study of Religion* 15:187-98.
_____. 1983. "Creative Productivity and Age: A Mathematical Model Based on a Two-Step Cognitive Process." *Developmental Review* 3:97-111.
_____. 1984. *Genius, Creativity, and Leadership*. Cambridge, MA: Harvard University Press.
_____. 1988. "Age and Outstanding Achievement: What Do We Know After a Century of Research?" *Psychological Bulletin* 104:251-67.
_____. 1990. "History, Chemistry, Psychology, and Genius: An Intellectual Autobiography of Historiometry." In *Theories of Creativity*, ed. Mark A. Runco and Robert S. Albert, pp. 92-115. Newbury Park, CA: Sage.
Snyder, Solomon H. 1989. *Brainstorming: The Science and Politics of Opiate Research*. Cambridge, MA: Harvard University Press.
Sovern, Michael I. 1989. "Higher Education: The Real Crisis." *New York Times Magazine*, January 22, pp. 24-25, 56.

Stephan, Paula E. 1976. "Human Capital Production: Life-Cycle Production with Different Learning Technologies." *Economic Inquiry* 14:539–57.
Stephan, Paula E., and Sharon G. Levin. 1987. *Demographic and Economic Determinants of Scientific Productivity*. Final Report to the National Science Foundation, the Exxon Educational Foundation, and the Alfred P. Sloan Foundation. Atlanta: Georgia State University.
_____. 1988. "Measures of Scientific Output and the Age-Productivity Relationship." In *Handbook of Quantitative Studies of Science and Technology*, ed. A. F. J. Van Raan, pp. 31–80. Amsterdam: Elsevier-North Holland.
_____. 1989. "The Effect of Cohort on the Publishing Productivity of U.S. Physicists." Paper presented to the annual meeting of the Society for the Social Studies of Science, Worcester, MA.
_____. 1990. "Scientific Rewards and the Economic Nature of the Scientific Endeavor." Paper presented to the annual meeting of the Society for the Social Studies of Science, Minneapolis.
_____. 1991. "Inequality in Scientific Performance: Adjustment for Inequality and Journal Impact." *Social Studies of Science* 21:351–68.
Stern, Richard E. 1990. "Uncitedness in the Biomedical Literature." *Journal of the American Society for Information Science* 41:193–6.
Stewart, John A. 1986. "Drifting Continents and Colliding Interests: A Quantitative Application of the Interests Perspective." *Social Studies of Science* 16:261–79.
Stewart, Walter W., and Ned Feder. 1987. "The Integrity of the Scientific Literature." *Nature* 325:207–14.
Survey of Doctorate Recipients (SDR). See National Research Council, various dates, *Science, Engineering, and Humanities Doctorates in the United States*.
Survey of Earned Doctorates (SED). See National Research Council, various dates, *Doctorate Recipients from United States Universities*. (Information from this survey is added to the Doctorate Records File.)
Szilagyi, Miklos, and Christopher Zell. 1990. Letter in *Physics Today* 43 (December): 92.
Teresi, Dick. 1990. "The Lone Ranger of Quantum Mechanics." *New York Book Review*, January 7, pp. 14–15.
Tobias, Sheila. 1990. *They're Not Dumb, They're Different: Stalking the Second Tier*. Tucson: Research Corporation.
Traweek, Sharon. 1988. *Beamtimes and Lifetimes: The World of High Energy Physicists*. Cambridge, MA: Harvard University Press.
Tuckman, Howard P. 1976. *Publication, Teaching and the Academic Reward Structure*. Lexington, MA: Lexington Books.
Tuckman, Howard P., and Jack Leahey. 1975. "What Is an Article Worth?" *Journal of Political Economy* 83:951–67.
Turner, Stephen P., and Darryl E. Chubin. 1979. "Chance and Eminence in Science: Ecclesiastes II." *Social Science Information* 18:437–49.
U.S. Congress, Office of Technology Assessment. 1989. *Higher Education for Science and Engineering: A Background Paper*. Washington, DC: U.S. Government Printing Office.
_____. 1991. *Federally Funded Research: Decisions for a Decade, Summary*. Washington, DC: U.S. Government Printing Office.
Vaughn, John C., and Robert M. Rosenzweig. 1990. "Heading Off a Ph.D. Shortage." *Issues in Science and Technology* 7:66–73.
Wasson, Tyler, ed. 1987. *Nobel Winners: An H. W. Wilson Biographical Dictionary*. Princeton, NJ: Visual Education Corporation.
Watson, James D. 1968. *The Double Helix*. New York: Atheneum.
Weber, Robert L. 1980. *Pioneers of Science: Nobel Prize Winners in Physics*. London: Institute of Physics.
Weinberg, Steven. 1974. "Unified Theories of Elementary-Particle Interaction." *Scientific American*, July, pp. 50–59.
Wolfe, Tom. 1983. "The Tinkerings of Robert Noyce." *Esquire*, December, pp. 346–74.
Wolpert, Lewis, and Alison Richards. 1988. *A Passion for Science*. New York: Oxford University Press.
Woodruff, Diana S. 1983. "Physiology and Behavior Relationships in Aging." In *Aging: Scientific Perspectives and Social Issues*, 2d ed., ed. Diane Woodruff and James E. Birren, pp. 178–201. Monterey, CA: Brooks/Cole.
Ziman, John M. 1976. *The Force of Knowledge*. Cambridge: Cambridge University Press.
Zuckerman, Harriet A. 1968. "Patterns of Name Ordering Among Authors of Scientific Papers: A Study of Social Symbolism and Its Ambiguity." *American Journal of Sociology* 74:276–91.

———. 1977. *The Scientific Elite*. New York: Free Press.
Zuckerman, Harriet, and Robert K. Merton. 1971. "Patterns of Evaluation in Science: Institutionalization, Structure and Functions of the Referee System." *Minerva* 9:66–100.
———. 1973. "Age, Aging, and Age Structure in Science." In *The Sociology of Science: Theoretical and Empirical Investigations*, ed. Norman Storer, pp. 497–575. Chicago: University of Chicago Press.

Index

Ability
  and age, 40
  as dimension of cognitive resources, 12–13
Acceptance. *See also* Resistance
  age and, 79, 81
  risk and, 79, 82–83
Adams, James, 86–87, 172n.29
Advantage, cumulative, 29–30
Agassiz, Louis, 75
Age. *See also* Older *entries*
  and acceptance, 81
  and cognitive resources, 37–41
  and cohort, and scientific productivity, PSR case study of, 132–55
  and creativity, 41–45, 48n.59
  importance of, 25–49
  and knowledge base, 37–38
  and lure of puzzle, 35–37
  measurement of remaining time by, 26–27, 32, 35, 46n.10
  and mental processes, 38–41
  "midpoint," 56, 73n.19
  productivity and, 50–74, 156–58
  and quest for gold, 33–35
  and quest for ribbon, 27–30
  and resistance, 75–77
  RPRT and, 4–5
  and will to do science, 27–37
Age and Achievement, 26, 72n.2
Age structure of science, 5–7, 10, 75–89, 157
  and economic growth, 84–88
  and rate of scientific discovery, 75–84
Aging effects, pure
  concept of, 133–34
  results of PSR case study, 140–45
Aging results, of case study, 140–45
Agricultural productivity, 86

AIDS, 20
Albert, Robert, 48n.60
Alfred P. Sloan Foundation, 74n.41
Allison, Paul, 30, 47n.19, 74n.37
Amabile, Teresa, 44
American Association for the Advancement of Science, 165
American Council on Education (ACE), 59, 153n.10
American Literature, 103
American Psychiatric Association, 38
American Review, 103
American scientific community, older, reasons for, 6–7, 84, 94–96, 110
Andrews, Frank, 61
Anomie, 37, 74n.38
Apollo space program, 78–79, 95
Apple computers, 14
Applied research, more, 167
Archimedes, 11–12
Aspen, 17
Astin, Alexander, 120–21, 121f., 122t.
Astin's selectivity index, 120–21, 130n.16&17
Atlanta Journal Constitution, 114n.49, 131n.40
Atomic and molecular physics, 140, 144, 149t., 151
Auburn University, 121
"Average" scientists, 57–72, 157
Avogadro's number, 57
Axelrod, Julius, 90, 113n.38, 158

Baby-boom and baby-bust generations, 116, 120, 127
Backwater(s), 45, 102, 103, 106, 108–9, 114n.68, 144, 151
Bacteriophages, 112n.1. *See also* "Phage group"

Ballard, Robert, 163
Baltimore, David, 163
Barash, David, 38
Bardeen, John, 3, 10n.3, 26, 69
Bayer, Alan, 60–61, 64, 71, 72, 144
BBC, 25
Beagle, 25
Beantimes and Lifetimes: The World of High Energy Physicists, 144
Beard, G. M., 42, 49n.68
Behavior, slowing of, 39
Belief system, 43
Bellcore, 19
Bell Laboratories, 3–4, 10n.3, 69
Berkeley, 21, 61–62, 163
"Best-fit" model, 60, 74n.35
Bethe, Hans, 107
Big Science, 85, 167–68
Biochemistry, 64, 74n.53
Biochemists, 15, 47n.19, 60, 67t., 71–72, 74n.37&53, 99, 138
Biologists, 43, 54, 99, 104
Birren, James, 39
Bishop, John, 121–22
Blau, Judith, 45
Bohr, Niels, 25
Book
  issues not covered in, 7
  plan of, 7–10
Boom to bust, job market cycle as, 94–96
Bowen, Howard, 125–26
Bowen, William, 130n.30, 170, 173n.48
Boyle, Robert, 14, 17
Boyle's law, 57
Braceland, Francis, 38
Bragg, Sir Lawrence, 25, 90
Brain
  age and, 38–39
  fads in research on, 104
Brain drain hypothesis, 117, 120, 123–26, 127
Brattain, Walter, 3–4, 10n.3, 69
Braun, Carl Ferdinand, 26
Brenner, Sydney, 15, 43
Broad, William J., 172n.37
Brookhaven National Laboratory, 69, 139
Bryn Mawr, 90
Burt, Cyril, 163

Calendar year, 135, 152n.1
California Institute of Technology, 120
Calvin College, 121
Cambridge, 16, 25, 90
Camelot Stores, 114n.49
Capital models, 34, 37–38, 44, 48n.36, 108, 145–46

Careers in science, loss of attractiveness of, 164
Carleton, 120
Carnegie Foundation for the Advancement of Teaching, 113n.31
Carson, Johnny, 19
Cattell, Raymond B., 48n.62
Cavendish Laboratory, 90
Census, U.S., 58
Center for European Nuclear Research (CERN), 139, 154n.23
Center for Marine Biology, 163
Change, in scientific knowledge
  coping with, by scientists, 106–9
  fads, 103–5, 146
  nature of, 102–3
  older scientists and, 77
  and RPRT, 158
  vintage and, 100
Chaos, 12, 14, 16, 17, 18, 22n.4, 23n.19
Chargaff's ratios, 57
Chemistry, 21, 53, 55t., 56, 59–60, 124, 131n.43
Chemists, 47n.19, 85, 74n.37
Chicago, University of, 120
Chronicle of Higher Education, 113n.28, 173n.41&45
Chu, Paul, 158, 168
Citations, 19, 21, 29, 47n.30, 57, 63, 122. *See also* Science Citation Index
  change and, 103
Clemson University, 121
Cliometrics, 81, 82
Cobweb model, 112n.15
Codification, 42, 105–6
Coe College, 121
Coggeshall, Porter, 153n.4
Cognitive resources
  age and, 37–41
  dimensions of, 11, 12–13
  keeping up and, 107
Cohen, Stanley, 12, 15
Cohort
  age and, and scientific productivity, PSR case study of, 132–55
  concept of, 115–16
  quality and, 116–27
Cohort differences, 160
Cohort results, in case study, 145–52
Cold fusion, 19, 22, 164
Cole, Jonathan, 29
Cole, Stephen, 29, 30, 47n.20, 54, 59–60, 61, 63, 122, 144
Colleagues, importance of, 15–17, 23n.23
Collecting data, 58–59
Columbia University, 99, 117

Index

Community, scientific, reasons for older American, 6–7
Compact disc, 103, 114n.49
Competition, increased
  for jobs and grants, 160–64
  for recognition, 53
Computers
  and change in science, 100
  dependence on, 14
  keeping up and, 107–8, 114n.65
Congress, U.S., 163, 168, 170, 173n.42
Consortium on Financing Higher Education (COFHE), 125
Continental drift, 81–82. *See also* Plate tectonic(s)
Coping, with change, by scientists, 106–9
Cornell University, 13, 120, 164
Cost(s)
  equipment, 14, 85
  indirect, 162, 171n.10
  opportunity, of research, 33–34, 37
Cox, Catherine, 54
Cray supercomputer, 11, 14
Creative options, for older scientists, 170
Creativity, 13. *See also* "Magic gland"
  age and, 41–45, 48n.59
  as process, 41–43, 48n.60&62, 49n.67&68
  as product, 43–45
Crick, Francis, 14, 15, 20, 24n.57, 25, 42, 45, 57, 90, 102, 103, 114n.50
Cross-sectional studies, 58, 59–61, 132–33, 134
Crowding, 117, 127
Culliton, Barbara, 172n.25
Cumulative (accumulative) advantage, 29–30, 47n.19, 92, 133, 157

Darsee, John, 163, 172n.26
Data, collecting of, 58–59
Darwin, Charles, 25, 42, 45, 57, 76, 81, 117
Demand, for scientists, 93–94, 170
Dennis, Wayne, 53, 54–55, 73n.6&17
Development. *See* Research and development
Diamond, Arthur, Jr., 21, 61–62, 81, 83
Dirac, P. A. M., 4, 25, 56, 57
Discovery, scientific. *See* Scientific discovery
Ditchdigging, 45, 79, 151
DNA, 14, 20, 24n.57, 42, 45, 57, 85, 90, 99, 100, 103, 162, 172n.30, 173n.46
Doctorate Records File (DRF), 74n.47, 112n.16&25, 113n.27, 120, 121f., 122t., 129n.15, 154n.20
Doing of science, 11–17
  age and will in, 27–37
  reasons for, 17–22

Donohue, Jerry, 16
Dupont, 69
Dutton, Jeffrey, 60–61, 64, 71, 72, 144
Dynamical Systems Collective, 16

Earliest works, mention of, 54
Earth, dynamic. *See* Plate tectonic(s)
Earth science(s), 21, 64, 91, 95, 105, 108, 138, 140, 151
  age-publishing profiles in, 141t.
  vintages for areas of, 148, 150
Earth scientists, 60, 70–71, 99
  normalized age-publishing profiles of, 66t., 70–71
Easterlin, Richard, 116, 117, 126, 129n.4, 126
Economic growth, science and, 84–88
Economic History Association, 81
The Economist, 112n.10
Effort, of scientists, 11–12
  age and, 34–35
Ehrenberg, Ronald, 113n.29
Einstein, Albert, 4, 25, 26, 36, 43, 45, 48n.42, 57, 105, 117, 139
Elaboration, of ideas, 42–43
Electroencephalogram (EEG), 39
Eltanin-19 profile, 20
Eminent scientists, 12, 26, 51, 57, 78, 122
  Lehman's studies of, 50–54
  Dennis's study of, 54–55
  Nobel laureates, studies of, 55–57
Employment distributions of physical scientists, 97–98, 97t., 113n.33
Engineering, 6, 21, 126, 130n.30
Equipment, importance of, 13–15, 23n.19
Erickson, Eric, 30
Esquire, 3
Evenson, Robert, 86
Excellent across the board, 169, 173n.42
Exxon Education Foundation, 74n.41

Faculty, expectations of, 169, 173n.43
Fads, in science, 104, 123, 146, 150
  importance of, 103–4
Fairchild Semiconductor, 22
Falk, G., 73n.18&21
Farnsworth, Philo Taylor, 42
Fashion. *See* Fads
Feder, Ned, 172n.26
Federally funded research and development centers (FFRDCs), 64, 69, 70, 91–92, 93, 97, 128, 136, 139, 142, 143, 144, 148
Fees, consulting and speaking, 21
Feigenbaum, Mitchell, 12
Fermi National Accelerator Laboratory, 69, 91, 139

Feynman, Richard, 18, 35, 164
Field, growth in, 98–99
Field theory, 99, 104, 113n.36, 150, 164
Fixed-effects model, 138, 153n.14, 155n.38
Fleming, Sir Alexander, 17, 57
Focus, shift of, 109
Ford, Joe, 17
Ford Mustang, 115
Fox, Mary Frank, 12
Franklin, Jon, 112n,5
Franklin, Rosalind, 14
Fraud, 163
Freeman, Richard, 22
Fresh point of view, 42, 43, 79
Freud, Sigmund, 42
Fulton, Oliver, 21
Fun, loss of, 164
Funding, 36, 161, 171n.8. *See also* Grants
   recommendations on, 167–68
   and risky research, 163, 171n.22
   stop-and-go, 165
   of young, 84
Future, recommendations for, 9–10, 165–66

Gale, Grant, 3
Galileo, 12, 14, 45
Gallo, Robert, 20
Galton, Sir Francis, 48n.50
Garfield, Eugene, 63
Gaston, Jerry, 19
Gauss, Karl, 25, 32
General Electric, 69
Generation. *See* Cohort
Genetics, 16, 18, 104, 108
Geologists, 55, 78–79
Geology, 60, 140, 143–44, 150t., 155n.41
Geophysics, 140, 144, 150t.
Georgia Institute of Technology, 17
German Review, 103
GeV ("giga"), 11
GI bill, 6
Giere, Ronald, 13, 16
Glashow-Weinberg-Salam unified theory, 154n.23
Gleick, James, 12, 16, 23n.19
GNP, 87
Goethe, 72n.1
Gold, 20–22, 27
Goldstein, Avram, 19
Good, Mary, 167, 172n.32
Gottlieb, Annie, 116
Gould, Stephen Jay, 22
Grading system, 124, 130n.31
Graduate Record Exam, 121

Grant(s), 36, 161, 171n.8. *See also* Funding
   versus creativity, 44
   first, 84
   and risky research, 163, 171n.22
Grassmuck, Karen, 173n.41
Great Depression, 92, 145
Gregory, Richard, 35
Grinnell College, 3, 100–101
Growth
   economic, science and, 84–88
   in field, 98–99
   of population of scientists, 53–54
   in scientific knowledge, 102–3
GULPs (grandiose, unnecessary, large projects), 168

Hagstrom, Warren, 18, 37
"Hardness of science," 105, 114n.61
Hargens, Lowell, 47n.11
Harmon, Lindsey, 12
Harré, Rom, 18
Harrington, David, 44
Hartnett, Rodney, 131n.43
Harvard, 20–21, 30, 106, 120, 125, 136, 154n.23, 162, 163, 168, 170, 173n.43
Hausman, Jerry, 153n.15
Hawking, Stephen, 105, 139
Hazard model, 82
Hazen, Robert, 12, 18, 19, 173n.45&46
Head start, 45
Heckman, James, 136, 153n.10, 153n.15
Heisenberg, Werner, 25
Heller, Scott, 113n.28
Hess, Harry, 16
Heterogeneous population, 30
High-energy physics. *See* Particle physics
Hilibrand, Lawrence E., 124
Hilts, Philip J., 172n.25
Hodgin, Dorothy, 26
"Holy Grail," 36, 105, 139, 144
Hooft, Gerard 't, 25
How science is done, 11–17
Hubble telescope, 85, 111, 162
Hull David, 16, 18, 20, 22, 81, 83, 105
Human capital models, 34, 37–38, 44, 108, 145–46
Human Genome Project, 85
Hurst, Robert E., 168, 173n.39
Huxley, Thomas, 78

IBM, 69
Ideation, 42–43
Identification, pooling and, 134–36
Illnesses, age-related, 41, 48n.58

Index

Imanishi-Kari, Thereza, 163, 171n.25
"Immediacy" of science, 103
Indiana University, 13, 90
Indirect costs, 162, 171n.10
Individual-specific fixed effects, 137–38, 153n.15&16&19
Institute of Scientific Information, 9, 47n.30
Institutions
 academic, study of, 59–61
 and access to equipment, 14–15, 23n.21
 job market for new Ph.D.'s at, 93
 resources of, 92
Intellectual climate, 91, 99–109
"Invisible college," 106, 114n.62
IQ, 12–13, 40, 41, 48n.54&60, 58, 119
Isis, 103

Jacob, François, 15
Jeffreys, Sir Harold, 75–76
Job market, 4–5, 83–84, 91–99, 128–29, 158, 159, 160
 as cohort effect, 145, 155n.35
 for new Ph.D.s, at research institutions, 93–94
 and Ph.D. cohorts, 136–37, 153n.11
Job market cycle, 94–96
Johns Hopkins, 90, 112n.3, 120
Jones, Lyle, 153n.4
Journals, 47n.11, 58. *See also* Publication
Journeymen scientists, 26, 32, 157
Julia sets, 12, 23n.7
Jung, Carl, 30

Keeping up, 106–8, 145–46
Kelvin, Lord, 75
Kepler, Johannes, 12
King, Julia, 172n.32
Kirby, Jack, 10n.1
Kislev, Yoav, 86
Knabenphysiks (boys physics), 25, 26
Knowledge, scientific
 change in. *See* Change
 growth in, 102–3
Knowledge base
 age and, 37–38
 as dimension of cognitive resources, 12
 vintage effects and, 146
 of younger scientists, 77
Knowledge effects, 86
Knowledge obsolescence, threat of, 101–2, 145–47
Krauze, Ted, 30, 47n.19, 74n.37
Kuhn, Thomas, 18, 79

Labor market. *See* Job market
Labor productivity, 89n.37&41
Laser(s), 85, 99, 100, 149t.
 and atomic and molecular physics, 140
 and change in science, 100
"Latest educated are best educated," 107, 132, 150, 151, 152
Lavoisier, Antoine Laurent, 83
Least publishable units (LPUs), 162
Lederberg, Joshua, 25
Lederman, Leon, 5–6, 44, 165, 166
Lehman, Harvey, 26, 50–51, 52t., 53–54, 53t., 55, 64, 72n.2, 73n.3,5&11, 84
Leibnitz, Gottfried Wilhelm von, 12
Levi-Montalcini, Rita, 15, 102
Levinson, Daniel, 31, 32, 36
Life science(s), 91, 95, 126
Life scientists, 98
Life span, of scientist, length of, 54
Life's passages, and will to do science, 30–33
Likelihood function, 139, 154n.18
Limited dependent variable, 138–39
Lindzey, Gardner, 153n.4
Little Science, Big Science, 117, 123
Location
 determinants of current, 96–98
 importance of, 91–92
 keeping up and, 107
Long, Scott, 30, 37n.19, 74n.37
Longitudinal studies, 58, 61–62, 133, 134, 135
Lorenz, Edward, 17, 23n.25
LPUs. *See* Least publishable units
Luria, Salvador, 13, 90, 112n.1

MaCurdy, Thomas, 153n.15
Maddala, G. S., 154n.16&19
"Magic gland," 13, 28, 30, 45. *See also* Creativity
The Making of a Scientist, 123
Manhattan Project, 6
Manniche, E., 73n.18&21
Mansfield, Edwin, 87
Many body theory, 100, 113n.36
Margin, persons from, 43, 44–45
Massachusetts Institute of Technology, 3, 124
Mathematicians, 18, 21, 61–62, 74n.38, 99
Mathematics, 21, 126
Matthew effect, 29, 47n.15
Maximum likelihood estimates, 154n.18
McCann, H. Gilman, 83
McClintock, Barbara, 57, 104, 108–9

Measurement
 of remaining time, by age, 26–27, 32, 35, 46n.10
 of scientific output, 57–58, 63–64, 135, 143t.
Medicine, 15, 55t., 56, 124, 125, 157
Meer, Simon van der, 154n.23
Memory, 39–40, 48n.52
Menard, Henry W., 98, 123
Mendel, Gregor, 14, 16, 18, 78
Mental processes, age and, 38–41
Merton, Robert, 18, 20, 29, 32, 42, 47n.11
Mervis, Jeffrey, 172n.20, 173n.49
Messeri, Peter, 82, 83, 88n.18
Methodology, in study of age, cohort, and scientific productivity, 134–39
Meyer, G. S., 122
Microbiology, 90
Mid-life Transition, 31
"Midpoint" age, 56, 73n.19
Mincer, Jacob, 107
Misconduct, 163–64
Model I, PSR study, 138, 154n.16&19
Model II, PSR study, 138, 154n.19
Mokyr, Joel, 117, 127
Molecular physicists, 140, 144
Molecular psychology, 90, 114n.5
Monetary rewards, 20–22, 27
Monod, Jacques, 15
Montagnier, Luc, 20
Morgan, T. H., 14
Motivation, as dimension of effort, 12
Multiple regression analysis, 137–38
Murrow, Edward R., 26
Mustang, 115

Napoleon, 18, 72n.1
National Academy of Sciences, 29, 62
National Aeronautics and Space Administration (NASA), 112n.18, 168
National Cancer Institute, 20
National Defense Education Act of 1958, 112n.18
National Institute of Mental Health, 48n.58
National Institutes of Health (NIH), 16, 71, 112n.18, 158, 163
National Merit scholars, 124, 130n.28&40, 151
National Research Council (NRC), 9, 62, 112n.16&17&25, 113n.30, 129n.13, 146, 152n.2, 170
National Science Board, 172n.32
National Science Foundation (NSF), 62, 74n.41, 112n.18, 131n.44, 161

Need, and opportunity, conjunction of, 170–71
Neugarten, Bernice, 31, 32, 47n.25
Newsweek, 19
Newton, Sir Isaac, 25, 42
New York Times, 26, 172n.25&37
NIH. *See* National Institutes of Health
Nixon, Richard, 95
Nobel Prize/laureate(s), 3, 4, 5, 9, 12, 13, 15, 17, 21, 25, 26, 33, 35, 44, 45, 69, 84, 90, 101, 104, 105, 107, 112n.1, 113n.38, 138, 154n.23, 156–57, 163, 164
 study of, 55–57, 55t.
Nose for success, 45
Noyce, Robert, 3, 6, 22, 26, 90, 158
NRC. *See* National Research Council
NSF. *See* National Science Foundation

Oakland University, 121
Obsolescence, knowledge, threat of, 101–2, 105–6, 145–47
Oceanography, 140, 143, 144, 150t.
Office of Scientific Integrity, 163
Office of Technology Assessment, 173n.42
Older American scientific community, reasons for, 6–7. *See also* Age *entries*
Older scientists, creative options for, 170
Olsen, Randall, 153n.10
On the Origin of the Species, 76, 81
OPEC, 88
Opiate receptor, 18, 21, 90, 101
Oppenheimer, J. Robert, 4, 26
Opportunity, need and, conjunction of, 170–71
Opportunity cost, of research, 33–34, 37
Ordinary least-squares regression (OLS) analysis, 138, 153n.10&11
Output. *See* Productivity
Own knowledge effects, 86

Particle physics, 14, 104, 105, 108, 109, 114n.51
 case study of, 139, 143, 144–45, 149t., 154n.20&24, 155n.31
Pascal, Blaise, 45
Passages, of life, 30–33
Passages, 30
Pasteur, Louis, 17
Pasteur Institute, 20
Pauli, Wolfgang, 25
Pauling, Linus, 4, 20, 42
Payment, for science, 166. *See also* Funding; Grants
Peace Corps, 170
Pelz, Donald, 61

Index 191

Persian Gulf War, 112n.10, 166
Personal computers (PCs), keeping up and, 107–8
Pert, Candace, 18, 19, 21, 90
"Phage group," 90, 112n.2, 114n.63
Ph.D.(s)
  employment distribution of physical scientists by, 97–98, 97t., 113n.33
  growth of, 94
  length of time since receipt of, 110–11, 110t.
  job market for new, at research institutions, 93
  median age of, 5
  recipients from the "most" selective schools, 121t., 122t.
Ph.D. cohorts, job market conditions and, 136–37, 153n.11
Ph.D.-to-population ratio, 118, 129n.13
Phi Beta Kappa, 125, 151
Philosophical reasons, for resistance, 77
Physical capital models, 34, 37–38, 48n.36
Physical Review Letters, 103
Physical sciences, 6, 126, 156–57
Physical scientists, employment distribution of, 97–98, 97t., 99, 113n.33
Physicists, 4, 21, 22, 29, 43, 91, 100, 101, 112n.13, 138
  atomic and molecular, 140, 144, 149t., 151
  normalized age-publishing profiles of, 65t., 69–70, 74n.51
Physics, 13, 21, 25, 26, 55t., 56, 60, 64, 95, 123, 124, 130n.30, 131n.43, 160
  age-publishing profiles in, 142t.
  condensed matter, 108, 109, 139, 144, 149t., 151
  particle. *See* Particle physics
  solid-state/condensed-matter, 139, 144
  vintages for areas of, 148, 149
Physiologists, 138
  normalized age-publishing profiles of, 68t., 71–72
Physiology, 64, 157
"Piltdown man," 163, 171n.23
Pitman, Walter, 20
Place, right. *See* Right place at the right time
Planck, Max, 26, 76, 157
Planck's constant, 57
Planck's principle, 76, 77–79, 157
  empirical investigation of, 80–84
Plate tectonic(s), 14, 16, 20, 23n.20, 25, 64, 80, 82, 99, 100, 108, 140
Policy, science, elements of rational, 166–70
Pooling, 133, 137
  and identification, 134–36

Poon, Leonard, 39–40
Population, of scientists
  growth in, 53–54
  heterogeneous, 30
Porter, Beverly, 22
Postdoctorate(s) 14, 16, 22, 96, 113n.26, 168
Premed, choice of, 124, 130n.35&40
Present value, 21, 24n.70, 33–34, 38n.35&38
Price, Derek de Solla, 13, 17, 47n.15, 102, 103, 113n.42, 117–18, 119t., 120, 122, 123, 129n.12
Priestly, Joseph, 75
Primary memory, 39
Princeton, 100, 158
Priority of discovery, 19–20, 24n.58
Prize money, 21
Product(s)
  creativity as, 43–45
  science and, 85, 87
Productivity. *See also* Economic growth; Publication
  age and, 50–74, 132–55, 156–58
  candidates for study of, 59
  cumulative advantage and, 29–30
  labor, 89n.37&41
  long-run consequences of impaired, 164–65, 172n.29&30
  measurement of, 57–58, 73n.24
  and salary, 21, 24n.71
  total or multifactor, 89n.37&41
Productivity growth, 86–89, 89n.37&41
Productivity of Scientists Research (PSR) Study, 62–72, 133–52, 156
  aging, results of, 140–45
  areas studied, 139–40
  cohort results of, 145–52
  data of, 62–68, 135
  methodology of, 134–39
Promotion, 35, 80
Pseudo R^2 statistic, 141, 143, 148, 154n.25, 155n.37
Publication, 28–29, 47n.11, 57–58, 59, 73n.23. *See also* Productivity
  by field and measure of output, 143t.
  in past five years, 103
  priority of discovery and, 19–20, 21, 24n.58
Pure aging effects, concept of, 133–34
Puzzle, 17, 18, 27
  age and lure of, 35–37

Quality, 9, 115–31, 159, 160
  and cohort, 116–27, 145, 151–52
Quantum (mechanics) theory, 4, 25, 85, 100, 151

Rabi, I. I., 4
Ramanujan, Srinivasa, 18
R&D. *See* Research and development
Rate of scientific discovery, age structure and, 75–84
Rational science policy, elements of, 166–70
Raup, David, 19
RCA, 69
Reading of papers, lack of, 37, 47n.30
Reassessment, 31–33, 35, 47n.25
Recognition, 17, 18–20. *See also* Ribbon
  eponymous, 57
  increase in competition for, 53
Reform, at elementary and secondary level, 169–70
Regression coefficients, 137–38
Reorganization, at universities, 168–69
Reputation. *See* Recognition
Reputational rank, 135, 153n.4
Research. *See also* Study
  basic, 86–88, 109
  more applied, 167
  values concerning importance of, 92
Research and development (R&D). *See also* Federally funded research and development centers (FFRDCs)
  academic, 95, 167
  in business and industry, 6, 86, 93, 128, 167
Research institutions. *See* Institutions
Research universities, classification of, 97, 113n.31
Resistance, age and, 75–77. *See also* Acceptance
Reskin, Barbara, 29
Resources
  cognitive. *See* Cognitive resources
  efficient allocation of, in science, 87
  of institutions, 92
  national, efficiency of use of, 85–86
Rhodes scholars, 125, 151
Ribbon, 18, 27. *See also* Recognition
  age and quest for, 27–30
Rigden, John S., 170, 173n.45&47
Right place at the right time (RPRT), 3–4, 8–9, 72, 111, 115, 116, 128, 132, 156, 158–59
  age and, 4–5
  concept of, 91
  field growth and, 98–99
  importance of, 5–6, 90–91, 111
  intellectual climate and, 99–101
  job market and, 91–92, 96–98
Risk, increased, for today's scientists, 109–11

Risky projects, 83, 162–63, 171n.20
RNA, 173n.46
Rockefeller University, 163
Roe, Anne, 12–13, 30, 123
Röentgen, Wilhelm Conrad, 26
Rosovsky, Henry, 20–21, 106
Royal Society of London's Catalog of Scientific Literature, 1800–1900, 54
Royalties, 21
RPRT. *See* Right place at the right time
R-square statistic, 60–61, 72, 141
  pseudo, 141, 143, 148, 154n.25, 155n.37
Rubbia, Carlo, 154n.23
Runco, Mark, 48n.60
Rutherford, Ernest, 75

"Sacred spark" hypothesis, 29–30, 47n.19
Sagan, Carl, 19
Salam, Abdus, 25–26, 104
Salary, productivity and, 21, 24n.71
Salomon Brothers, 124
Salthouse, Timothy, 48n.50
Sample-selectivity bias, 136–37
Saudi Arabia, 112n.10
Schaie, Warner, 40
Scherer, Frederic, 86
Scholastic Aptitude Test (SAT) scores, 127, 131n.43
Schrödinger, Erwin, 22, 26
Schuster, Jack, 125–26
Science
  age structure of. *See* Age structure
  doing of. *See* Doing of science
  earth. *See* Earth science(s)
  and economic growth, 84–88
  education, reform of, 169–70
  fads in, importance of, 103–4
  golden age for, 94–95
  "hardness" of, 105, 114n.61
  quality in. *See* Quality
  secularization of, 117, 123
Science, 20, 81, 84
Science Citation Index (SCI), 9, 47n.30, 63, 73n.23, 74n.45, 153n.3
Science policy, elements of rational, 166–70
"Science: The End of the Frontier?," 165
Scientific community, reasons for older American, 6–7
Scientific discovery
  chance and, 17
  priority of, 19–20, 24n.58
  rate of, age structure and, 75–84
  youth and, 25–26, 45, 157
Scientific knowledge. *See* Knowledge, scientific

Index

Scientific output. *See* Productivity
Scientific productivity. *See* Productivity
Scientist(s)
   "average," 57–72
   contributions of, 11–13
   coping of, with change, 106–9
   eminent, 50–57
   growth in population of, 53–54
   heterogeneous population of, 30
   life span of, length of, 54
   older, creative options for, 170
   physical, employment distributions of, 97–98, 97t., 113n.33
   supply and demand for, 93–94, 170
   of today, increased risk for, 109–11
The Scientist, 172n.20&32, 173n.49
SDR. *See* Survey of Doctorate Recipients
Secondary memory, 39
Secularization of science, 117, 123
Selectivity-bias correction variable, 136, 153n.10
Selectivity hypothesis, 117–22, 127
Sensory memory, 39
Serendipity, 17, 23n.33
Shaw, Robert Stetson, 16
Sheehy, Gail, 30
Shift of focus, 109
Shockley, William, 10n.3, 69
Sidewaters, 108–9
Simonton, Dean Keith, 42–43, 48n.59&62, 49n.67&68, 53, 54, 117, 127
Single-investigator projects (SIPs), 161, 168
Sixties generation, 116
"Skimming the cream," 118–20
Sloppiness, 163–64, 171n.26
"Small" science, 167–68, 172n.34
S-matrix (scattering matrix) theory, 104, 150, 155n.42
Snyder, Solomon, 18, 19, 21, 24n.66, 90, 101, 113n.38, 114n.3
Sociocultural hypothesis, 117, 126–27
Sociological reasons, for resistance, 77–79
Solid-state/condensed-matter physics, 139, 144
"Sophistication inflation," 162
Sosa, Julie, 130n.30, 170, 173n.48
Source Index, of *SCI*, 74n.45
Sovern, Michael, 117
Space station, 85
"Spillover" effects, 86
Sputnik, 6, 94, 124
Stanford, 120
Stanford Linear Accelerator, 22, 69, 139
Stewart, John, 30, 47n.19, 82
Stewart, Walter, 172n.26

Stop-and-go funding, 165
Storer, Norman, 114n.61
Stout, Jim, 14
String theory, 155n.42
Structure, age. *See* Age structure
Studies in English Literature, 103
Study, of productivity, candidates for, 59. *See also* Research
Success, nose for, 45
Superconducting Supercollider, 14, 69, 85, 166
Superconductivity, 16, 19, 23n.5, 26, 168
Superconductor(s), 12, 18, 19, 22, 23n.5, 26, 139
   high-temperature, 12, 18, 22, 23n.5, 139
Supply and demand for scientists, 93–94
Survey of Doctorate Recipients (SDR), 62–63, 64, 97t., 110t., 113n.27, 135, 137, 152n.2, 153n.3,4&11, 154n.20, 171n.1
Survey of Earned Doctorates, 9, 113n.26&27
Szilard, Leo, 43

Tabletop research, 14, 16, 167
Talent deficit, 151
Taylor, William, 153n.15
Teaching, recommendations regarding, 169, 173n.41
Teams, 15–16
Tenure, 35, 47n.39, 80, 83–84, 136, 160, 169, 173n.43
Teresi, Dick, 22
Tertiary memory, 48n.52
Tessner, Peter, 81, 83
Time
   to be creative, 44
   as cost of research, 33
   as dimension of effort, 11–12
   keeping up and, 107
   length of, since Ph.D. received, 110–11, 110t.
   measurement of remaining, by age, 26–27, 32, 35, 46n.10
   right. *See* Right place at the right time
"Time-period effects," 133, 134–35
Tobias, Sheila, 173n.45
Tobin, James, 138–39
Tobit, 138–39, 141, 143, 148, 154n.19
Total or multifactor productivity, 89n.37&41
"Transhistorical time series analysis," 127
Transistors, and change in science, 100
Traweek, Sharon, 112n.13, 123, 130n.26, 144
Trefill, James, 173n.45&46
Trivial pursuits, 80, 88n.18, 84
Trow, Martin, 21
Tuckman, Howard, 21

Union Carbide, 69
U.S. Geological Survey, 59
Unification theory, 105, 150, 155n.42
Universities
  classification of, 97, 113n.31
  reorganization at, 168–69
University of Arkansas, 121
University of California at Berkeley, 21, 61–62, 163
University of Delaware, 90
University of Minnesota, 14
University of Missouri at Columbia, 121
University of Oklahoma, 173n.39
University of Oregon, 121
Upgrading, 108, 145–46

Vanderbilt, 120
Vassar, 120
Vietnam War, 6, 84, 94, 112n.10, 116
Vine, Fred, 16, 20, 25
Vine-Matthews hypothesis, 25
Vintage(s), 5, 128, 129, 159
  as cohort effect, 145–48, 149t., 150–52, 150t.
  concept of, 99–101, 102
  keeping up and, 107–8

and obsolescence, threat of, 101–2
quality and, 115

Watson, James, 4, 13, 14, 16, 17, 20, 24n.57, 25, 26, 42, 45, 57, 90, 103, 114n.50, 138, 162
Wegener, Alfred, 81
Weinberg, Steven, 104, 105
Westinghouse, 69
Why science is done, 17–22
Wilkins, Maurice, 17
Will to do science, age and, 27–37
Wilson, Kenneth, 13
Wilson, J. Tuzo, 16
Within estimator, 154n.16
Wolpert, Lewis, 78
Woods Hole Laboratory, 78, 163
World War II, 6, 10n.3, 25, 88
"W particle," 154n.23
Wrong place at the wrong time, 92, 112n.10

Yale, 32, 120, 124
Yandel, Gerry, 114n.49

Zuckerman, Harriet, 32, 42, 45, 47n.11, 56–57, 73n.18&21